Hydraulics and Pneumatics

Hydraulics and Pneumatics
A technician's and engineer's guide

Second edition

Andrew Parr *MSc., CEng., MIEE, MInstMC*

BUTTERWORTH
HEINEMANN

OXFORD BOSTON JOHANNESBURG MELBOURNE NEW DELHI SINGAPORE

Butterworth-Heinemann is an imprint of Elsevier
Linacre House, Jordan Hill, Oxford OX2 8DP, UK
30 Corporate Drive, Suite 400, Burlington, MA 01803, USA

First edition 1991
Reprinted 1992, 1993, 1995
Second edition 1998
Paperback edition 1999
Reprinted 2000 (twice), 2002, 2003, 2004, 2005 (twice), 2006, 2007

Notice
No responsibility is assumed by the publisher for any injury and/or damage to persons
or property as a matter of products liability, negligence or otherwise, or from any use
or operation of any methods, products, instructions or ideas contained in the material
herein. Because of rapid advances in the medical sciences, in particular, independent
verification of diagnoses and drug dosages should be made

British Library Cataloguing in Publication Data
Parr, E. A. (Eric Andrew)
  Hydraulics and pneumatics: a technician's and engineer's guide. – 2nd ed.
  1. Hydraulics 2. Hydraulic engineering 3. Pneumatics
  I. Title
  621.5'1

Library of Congress Cataloging-in-Publication Data
Parr, E. A. (E. Andrew)
  Hydraulics and pneumatics: a technician's and engineer's guide
  Andrew Parr. 2nd ed.
  p. cm.
  Includes index
  1. Hydraulic machinery – Handbooks, manuals, etc. 2. Pneumatic
  Machinery – Handbooks, manuals, etc. I. Title
  TJ840.P27   98-22010
  621.5'1–dc21   CTP

ISBN–13: 978-0-7506-4419-9
ISBN–10: 0-7506-4419-2

For information on all Butteraworth-Heinemann publications
visit our website at books.elsevier.com

Printed and bound in *Great Britain*

07 08 09 10 10

Working together to grow
libraries in developing countries

www.elsevier.com | www.bookaid.org | www.sabre.org

ELSEVIER   BOOK AID   Sabre Foundation
             International

Contents

Preface

Machines should work, people should think
The IBM Pollyanna Principle

Practically every industrial process requires objects to be moved, manipulated or be subjected to some form of force. This is generally accomplished by means of electrical equipment (such as motors or solenoids), or via devices driven by air (pneumatics) or liquids (hydraulics).

Traditionally, pneumatics and hydraulics are thought to be a mechanical engineer's subject (and are generally taught as such in colleges). In practice, techniques (and, more important, the fault-finding methodology) tend to be more akin to the ideas used in electronics and process control.

This book has been written by a process control engineer as a guide to the operation of hydraulic and pneumatics systems. It is intended for engineers and technicians who wish to have an insight into the components and operation of a pneumatic or hydraulic system. The mathematical content has been deliberately kept simple with the aim of making the book readable rather than rigorous. It is not, therefore, a design manual and topics such as sizing of pipes and valves have been deliberately omitted.

This second edition has been updated to include recent developments such as the increasing use of proportional valves, and includes an expanded section on industrial safety.

Andrew Parr
Isle of Sheppey
ea_parr@compuserve.com

1

Fundamental principles

Industrial prime movers

Most industrial processes require objects or substances to be moved from one location to another, or a force to be applied to hold, shape or compress a product. Such activities are performed by Prime Movers; the workhorses of manufacturing industries.

In many locations all prime movers are electrical. Rotary motions can be provided by simple motors, and linear motion can be obtained from rotary motion by devices such as screw jacks or rack and pinions. Where a pure force or a short linear stroke is required a solenoid may be used (although there are limits to the force that can be obtained by this means).

Electrical devices are not, however, the only means of providing prime movers. Enclosed fluids (both liquids and gases) can also be used to convey energy from one location to another and, consequently, to produce rotary or linear motion or apply a force. Fluid-based systems using liquids as transmission media are called hydraulic systems (from the Greek words *hydra* for water and *aulos* for a pipe; descriptions which imply fluids are water although oils are more commonly used). Gas-based systems are called Pneumatic systems (from the Greek *pneumn* for wind or breath). The most common gas is simply compressed air. although nitrogen is occasionally used.

The main advantages and disadvantages of pneumatic or hydraulic systems both arise out of the different characteristics of low density compressible gases and (relatively) high density incompressible liquids. A pneumatic system, for example, tends to have a 'softer' action than a hydraulic system which can be prone to producing noisy and wear inducing shocks in the piping. A liquid-based hydraulic system, however, can operate at far higher pressures than a pneumatic system and, consequently, can be used to provide very large forces.

To compare the various advantages and disadvantages of electrical pneumatic and hydraulic systems, the following three sections consider how a simple lifting task could be handled by each.

A brief system comparison

The task considered is how to lift a load by a distance of about 500 mm. Such tasks are common in manufacturing industries.

An electrical system

With an electrical system we have three basic choices; a solenoid, a DC motor or the ubiquitous workhorse of industry, the AC induction motor. Of these, the solenoid produces a linear stroke directly but its stroke is normally limited to a maximum distance of around 100 mm.

Both DC and AC motors are rotary devices and their outputs need to be converted to linear motion by mechanical devices such as wormscrews or rack and pinions. This presents no real problems; commercial devices are available comprising motor and screw.

The choice of motor depends largely on the speed control requirements. A DC motor fitted with a tacho and driven by a thyristor drive can give excellent speed control, but has high maintenance requirements for brushes and commutator.

An AC motor is virtually maintenance free, but is essentially a fixed speed device (with speed being determined by number of poles and the supply frequency). Speed can be adjusted with a variable frequency drive, but care needs to be taken to avoid overheating as most motors are cooled by an internal fan connected directly to the motor shaft. We will assume a fixed speed raise/lower is required, so an AC motor driving a screwjack would seem to be the logical choice.

Neither type of motor can be allowed to stall against an end of travel stop, (this is not quite true; specially-designed DC motors, featuring good current control on a thyristor drive together with an external cooling fan, *can* be allowed to stall), so end of travel limits are needed to stop the drive.

We have thus ended up with the system shown in Figure 1.1 comprising a mechanical jack driven by an AC motor controlled by a reversing starter. Auxiliary equipment comprises two limit switches, and a motor overload protection device. There is no practical load limitation provided screw/gearbox ratio, motor size and contactor rating are correctly calculated.

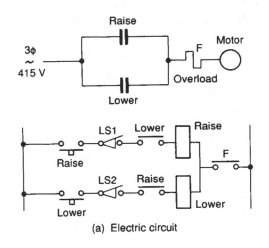

(a) Electric circuit

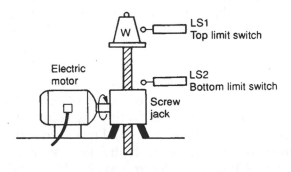

(b) Physical layout

Figure 1.1 *Electrical solution, based on three phase motor*

A hydraulic system

A solution along hydraulic lines is shown in Figure 1.2. A hydraulic linear actuator suitable for this application is the ram, shown schematically in Figure 1.2a. This consists of a movable piston connected directly to the output shaft. If fluid is pumped into pipe A the piston will move up and the shaft will extend; if fluid is pumped into pipe B, the shaft will retract. Obviously some method of retrieving fluid from the non-pressurised side of the piston must be incorporated.

The maximum force available from the cylinder depends on fluid pressure and cross sectional area of the piston. This is discussed further in a later section but, as an example, a typical hydraulic pressure of 150 bar will lift 150 kg cm^{-2} of piston area. A load of 2000 kg could thus be lifted by a 4.2cm diameter piston.

A suitable hydraulic system is shown in Figure 1.2b. The system requires a liquid fluid to operate; expensive and messy and, consequently, the piping must act as a closed loop, with fluid transferred from a storage tank to one side of the piston, and returned from the other side of the piston to the tank. Fluid is drawn from the tank by a pump which produces fluid flow at the required 150 bar. Such high pressure pumps, however, cannot operate into a dead-end load as they deliver constant volumes of fluid from input to output ports for each revolution of the pump shaft. With a dead-end load, fluid pressure rises indefinitely, until a pipe or the pump itself fails. Some form of pressure regulation, as shown, is therefore required to spill excess fluid back to the tank.

Cylinder movement is controlled by a three position changeover valve. To extend the cylinder, port A is connected to the pressure line and port B to the tank. To reverse the motion, port B is connected to the pressure line and port A to the tank. In its centre position the valve locks the fluid into the cylinder (thereby holding it in position) and dead-ends the fluid lines (causing all the pump output fluid to return to the tank via the pressure regulator).

There are a few auxiliary points worthy of comment. First, speed control is easily achieved by regulating the volume flow rate to the cylinder (discussed in a later section). Precise control at low speeds is one of the main advantages of hydraulic systems.

Second, travel limits are determined by the cylinder stroke and cylinders, generally, can be allowed to stall at the ends of travel so no overtravel protection is required.

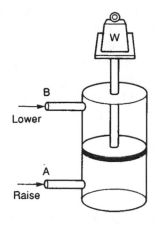

(a) Hydraulic cylinder

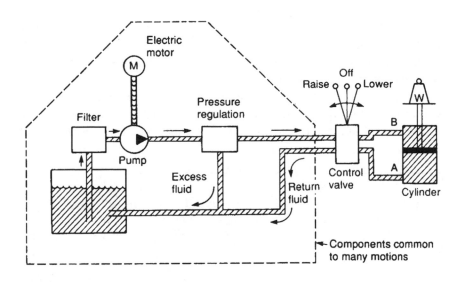

(b) Physical components

Figure 1.2 *Hydraulic solution*

Third, the pump needs to be turned by an external power source; almost certainly an AC induction motor which, in turn, requires a motor starter and overload protection.

Fourth, hydraulic fluid needs to be very clean, hence a filter is needed (shown in Figure 1.2b) to remove dirt particles before the fluid passes from the tank to the pump.

One final point worth mentioning is that leaks of fluid from the system are unsightly, slippery (hence hazardous) and environmentally very undesirable A major failure can be catastrophic.

At first sight Figure 1.2b appears inordinately complicated compared with the electrical system of Figure 1.1, but it should be remembered all parts enclosed in the broken-lined box in Figure 1.2 are common to an area of plant and not usually devoted to just one motion as we have drawn.

A pneumatic system

Figure 1.3 shows the components of a pneumatic system. The basic actuator is again a cylinder, with maximum force on the shaft being determined by air pressure and piston cross sectional area. Operating pressures in pneumatic systems are generally much lower than those in a hydraulic systems; 10 bar being typical which will lift 10 kg cm^{-2} of piston area, so a 16 cm diameter piston is required to lift the 2000 kg load specified in the previous section. Pneumatic systems therefore require larger actuators than hydraulic systems for the same load.

The valve delivering air to the cylinder operates in a similar way to its hydraulic equivalent. One notable difference arises out of the simple fact that air is free; return air is simply vented to atmosphere.

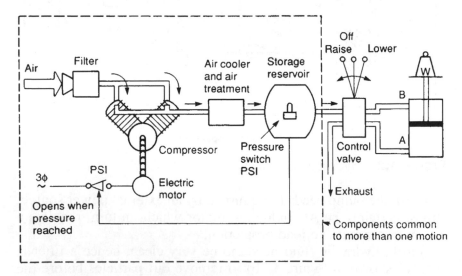

Figure 1.3 *Pneumatic solution*

Air is drawn from the atmosphere via an air filter and raised to required pressure by an air compressor (usually driven by an AC motor). The air temperature is raised considerably by this compressor. Air also contains a significant amount of water vapour. Before the air can be used it must be cooled, and this results in the formation of condensation So, the air compressor must be followed by a cooler and air treatment unit.

Compressibility of a gas makes it necessary to store a volume of pressurised gas in a reservoir, to be drawn on by the load. Without this reservoir, a slow exponential rise of pressure results in a similar slow cylinder movement when the valve is first opened. The air treatment unit is thus followed by an air reservoir.

Hydraulic systems require a pressure regulator to spill excess fluid back to the tank, but pressure control in a hydraulic system is much simpler. A pressure switch, fitted to the air reservoir, starts the compressor motor when pressure falls and stops it again when pressure reaches the required level.

The general impression is again one of complexity, but units in the broken-lined box are again common to one plant or even a whole site. Many factories produce compressed air at one central station and distribute an air ring main to all places on the site in a similar way to other services such as electricity, water or gas.

A comparison

Table 1.1 gives superficial comparisons of the various systems discussed in the previous sections.

Definition of terms

There is an almost universal lack of standardisation of units used for measurement in industry, and every engineer will tell tales of gauges indicating, say, velocity in furlongs per fortnight. Hydraulics and pneumatic systems suffer particularly from this characteristic, and it is by no means unusual to find pressure indicated at different locations in the same system in bar, kpascal and psi.

There is, however, a welcome (and overdue) movement to standardisation on the International System (SI) of units, but it will be some time before this is complete. The engineer will therefore encounter many odd-ball systems in the years to come.

Table 1.1   Comparisons of electrical, hydraulic and pneumatic systems

	Electrical	*Hydraulic*	*Pneumatic*
Energy source	Usually from outside supplier	Electric motor or diesel driven	Electric motor or diesel driven
Energy storage	Limited (batteries)	Limited (accumulator)	Good (reservoir)
Distribution system	Excellent, with minimal loss	Limited basically a local facility	Good. can be treated as a plant wide service
Energy cost	Lowest	Medium	Highest
Rotary actuators	AC & DC motors. Good control on DC motors. AC motors cheap	Low speed. Good control. Can be stalled	Wide speed range. Accurate speed control difficult
Linear actuator	Short motion via solenoid. Otherwise via mechanical conversion	Cylinders. Very high force	Cylinders. Medium force
Controllable force	Possible with solenoid & DC motors Complicated by need for cooling	Controllable high force	Controllable medium force
Points to note	Danger from electric shock	Leakage dangerous and unsightly. Fire hazard	Noise

Any measurement system requires definition of the six units used to measure:

- length:
- mass;
- time;
- temperature;
- electrical current;
- light intensity.

Of these, hydraulic/pneumatic engineers are primarily concerned with the first three. Other units (such as velocity, force, pressure) can be defined in terms of these basic units. Velocity, for example, is defined in terms of length/time.

The old British Imperial system used units of foot, pound and second (and was consequently known as the *fps system*). Early metric systems used centimetre, gramme and second (known as the *cgs system*), and metre, kilogramme and second (the *mks system*). The mks system evolved into the *SI system* which introduces a more logical method of defining force and pressure (discussed in later sections). Table 1.2 gives conversions between basic simple units.

Table 1.2   Fundamental mechanical units

Mass
   1 kg = 2.2046 pound (lb) = 1000 gm
   1 lb = 0.4536 kg
   1 ton (imperial) = 2240 lb = 1016 kg = 1.12 ton (US)
   1 tonne = 1000 kg = 2204.6 lb = 0.9842 ton (imperial)
   1 ton (US) = 0.8929 ton (imperial)

Length
   1 metre = 3.281 foot (ft) = 1000 mm = 100 cm
   1 inch = 25.4 mm = 2.54 cm
   1 yard = 0.9144 m

Volume
   1 litre = 0.2200 gallon (imperial) = 0.2642 gallon (US)
   1 gallon (imperial) = 4.546 litre = 1.2011 gallon (US)
      = 0.161 cubic ft
   1 gallon (US) = 3.785 litre = 0.8326 gallon (imperial)
   1 cubic meter = 220 gallon (imperial) = 35.315 cubic feet
   1 cubic inch = 16.387 cubic centimetres

Mass and force

Pneumatic and hydraulic systems generally rely on pressure in a fluid. Before we can discuss definitions of pressure, though, we must first be clear what is meant by everyday terms such as weight, mass and force.

We all are used to the idea of weight, which is a *force* arising from gravitational attraction between the mass of an object and the earth. The author weighs 75 kg on the bathroom scales; this is equivalent to saying there is 75 kg *force* between his feet and the ground.

Weight therefore depends on the force of gravity. On the moon, where gravity is about one sixth that on earth, the author's weight would be about 12.5 kg; in free fall the weight would be zero. In all cases, though, the author's *mass* is constant.

The British Imperial fps system and the early metric systems link mass and weight (force) by defining the unit of force to be the gravitational attraction of unit mass at the surface of the earth. We thus have a mass defined in pounds and force defined in pounds force (lbs f) in the fps system, and mass in kilogrammes and force in kg f in the mks system.

Strictly speaking, therefore, bathroom scales which read 75 kg are measuring 75 kg f, not the author's mass. On the moon they would read 12.5 kg f, and in free fall they would read zero.

If a force is applied to a mass, acceleration (or deceleration) will result as given by the well known formula:

$$F = ma. \tag{1.1}$$

Care must be taken with units when a force F is defined in lbs f or kg f and mass is defined in lbs or kg, because resulting accelerations are in units of g; acceleration due to gravity. A force of 25 kg f applied to the author's mass of 75 kg produces an acceleration of 0.333 g.

The SI unit of force, the newton (N), is defined not from earth's gravity, but directly from expression 1.1. A newton is defined as the force which produces an acceleration of 1 m s^{-2} when applied to a mass of 1 kg.

One kgf produces an acceleration of 1 g (9.81 ms^{-2}) when applied to a mass of 1 kg. One newton produces an acceleration of 1 ms^{-2} when applied to mass of 1 kg. It therefore follows that:

$$1 \text{ kg f} = 9.81 \text{ N}$$

but as most instruments on industrial systems are at best 2% accurate it is reasonable (and much simpler) to use:

$$1 \text{ kg f} = 10 \text{ N}$$

for practical applications.

Table 1.3 gives conversions between various units of force.

Table 1.3   Units of force

---

1 newton (N) = 0.2248 pound force (lb f)
    = 0.1019 kilogram force (kg f)
1 lb f = 4.448N = 0.4534 kg f
1 kg f = 9.81N = 2.205 lb
Other units are
    dynes (cgs unit); $1 \text{ N} = 10^5$ dynes
    ponds (gram force); 1 N = 102 ponds
SI unit is the newton:
    $N = \text{kg ms}^{-2}$

Pressure

Pressure occurs in a fluid when it is subjected to a force. In Figure 1.4 a force F is applied to an enclosed fluid via a piston of area A. This results in a pressure P in the fluid. Obviously increasing the force increases the pressure in direct proportion. Less obviously, though, decreasing piston area also increases pressure. Pressure in the fluid can therefore be defined as the force acting per unit area, or:

$$P = \frac{F}{A}. \tag{1.2}$$

Although expression 1.2 is very simple, there are many different units of pressure in common use. In the Imperial fps system. for example, F is given in lbs f and A is given in square inches to give pressure measured in pound force per square inch (psi).

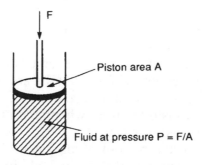

Figure 1.4   *Pressure in a fluid subjected to a force*

In metric systems, F is usually given in kgf and A in square centimetres to give pressure in kilogram/force per square centimetre (kg f cm^{-2}).

The SI system defines pressure as the force in newtons per square metre (N m^{-2}). The SI unit of pressure is the pascal (with 1 Pa = 1 N m^{-2}). One pascal is a very low pressure for practical use, however, so the kilopascal (1 kPa = 10^{3}Pa) or the megapascal (1 MPa = 10^6 Pa) are more commonly used.

Pressure can also arise in a fluid from the weight of a fluid. This is usually known as the head pressure and depends on the height of fluid. In Figure 1.5 the pressure at the bottom of the fluid is directly proportional to height h.

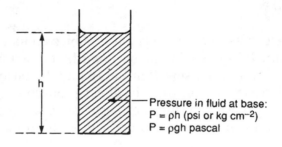

Figure 1.5   *Head pressure in a fluid*

In the Imperial and metric systems head pressure is given by:

$$P = \rho h. \tag{1.3}$$

where ρ is the density and h the height (both in the correct units) to give P in psi or kg cm^{-2}.

In the SI system expression 1.3. is re-arranged as:

$$P = \rho gh. \tag{1.4}$$

where g is the acceleration due to gravity (9.81 ms^{-2}) to give the pressure in pascal.

Pressure in a fluid can, however, be defined in terms of the *equivalent* head pressure. Common units are millimetres of mercury and centimetres, inches, feet or metres of water. The suffix *wg* (for *water gauge*) is often used when pressure is defined in terms of an equivalent head of water.

We live at the bottom of an ocean of air, and are consequently subject to a substantial pressure head from the weight of air above us. This pressure, some 15 psi, 1.05 kg f cm^{-2}, or 101 kPa, is called an atmosphere, and is sometimes used as a unit of pressure.

It will be noted that 100 kPa is, for practical purposes, one atmosphere As this is a convenient unit for many applications 100 kPa (10^5 Pa or 0.1 MPa) has been given the name *bar*. Within the accuracy of instrumentation generally found in industry 1 bar = 1 atmosphere.

There are three distinct ways in which pressure is measured, shown in Figure 1.6. Almost all pressure transducers or transmitters measure the pressure *difference* between two input ports. This is known as *differential pressure,* and the pressure transmitter in Figure 1.6a indicates a pressure of P_1-P_2.

In Figure 1.6b the low pressure input port is open to atmosphere, so the pressure transmitter indicates pressure above atmospheric pressure. This is known as *gauge pressure,* and is usually denoted by a *g* suffix (e.g. psig). Gauge pressure measurement is almost universally used in hydraulic and pneumatic systems (and has been implicitly assumed in all previous discussions in this chapter).

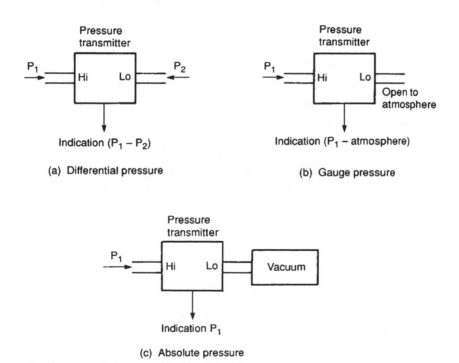

(a) Differential pressure

(b) Gauge pressure

(c) Absolute pressure

Figure 1.6   *Different forms of pressure measurement*

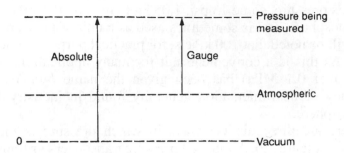

Figure 1.7   *Relationship between absolute and gauge pressures*

Figure 1.6c shows the pressure transmitter measuring pressure with respect to a vacuum. This is known as *absolute pressure* and is of importance when the compression of gases is considered. The relationship between absolute and gauge pressure is illustrated in Figure 1.7. Pressure measurement and gas compression are discussed in later sections. Table 1.4 compares units of pressure. A typical hydraulic system operates at 150 bar, while typical pneumatic systems operate at 10 bar.

Work, energy and power

Work is done (or energy is transferred) when an object is moved against a force, and is defined as:

$$\text{work} = \text{force} \times \text{distance moved.} \tag{1.5}$$

In the Imperial fps system expression 1.5 gives a unit of ft lb f. For metric systems the unit is cm kg f. The SI unit of work is the joule, where $1\ J = 1\ N\ m\ (= 1\ m^2\ kg\ s^{-2})$. Table 1.5 compares these, and other, units of work.

Power is the rate at which work is performed:

$$\text{power} = \frac{\text{work}}{\text{time}}. \tag{1.6}$$

The SI unit of power is the watt, defined as $1\ J\ s^{-1}$. This is by far the most common unit of power, as it is almost universally used for the measurement of electrical power.

The Imperial system uses horse power (Hp) which was used historically to define motor powers. One horse power is defined as $550\ \text{ft lb f s}^{-1}$. Table 1.6 compares units of power.

Table 1.4   Units of pressure

1 bar = 100 kPa
     = 14.5 psi
     = 750 mmHg
     = 401.8 inches W G
     = 1.0197 kgf cm^{-2}
     = 0.9872 atmosphere
1 kilopascal = 1000 Pa
          = 0.01 bar
          = 0.145 psi
          = 1.0197 × 10^{-3} kgf cm^{-2}
          = 4.018 inches W G
          = 9.872 × 10^{-3} atmosphere
1 pound per square inch (psi) = 6.895 kPa
                      = 0.0703 kgf cm^{-2}
                      = 27.7 inches W G
1 kilogram force per square cm (kgf cm-2) = 98.07 kPa
                                = 14.223 psi
1 Atmosphere = 1.013 bar
          = 14.7 psi
          = 1.033 kgf cm^{-2}
SI unit of pressure is the pascal (Pa) lPa = lN m^{-2}
Practical units are the bar and the psi.

Table 1.5   Units of work (energy)

1 joule (J) = 2.788 x 10^{-4} Wh (2.788 × 10^{-7} kWh)
        = 0.7376 ft lbf
        = 0.2388 calories
        = 9.487 × 10^{-4} British thermal units (BTu)
        = 0.102 kgf m
        = 10^{7} ergs (cgs unit)
SI unit of work is the joule (J)
lJ = 1 N m
  = 1 m^2 kg s^{-2}

Table 1.6   Units of power

---

  1 kwatt (kw) = 1.34 Hp
            = 1.36 metric Hp
            = 102 kgf m s^{-1}
            = 1000 W
1 horse power (Hp) = 0.7457 kw
                 = 550 Ft lb s^{-1}
                 = 2545 BTU h^{-1}
SI unit of power (and the practical unit) is the watt (W)

Work can be considered as the time integral of power (often described loosely as *total power used*). *As* electrical power is measured in watts or kilowatts (1 kW= 10^3W), the kilowatt hour (kW h) is another representation of work or energy.

Torque

The term *torque is* used to define a rotary force. and is simply the product of the force and the effective radius as shown in Figure 1.8. We thus have:

$$T = F \times d. \tag{1.7}$$

In the Imperial system the unit is lbf ft, in metric systems the unit is kgf m or kgf cm, and in SI the unit is N m.

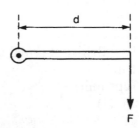

Figure 1.8   *Definition of torque*

Pascal's law

Pressure in an enclosed fluid can be considered uniform throughout a practical system. There may be small differences arising from head pressures at different heights, but these will generally be negligible compared with the system operating pressure. This equality of pressure is known as *Pascal's law,* and is illustrated in Figure 1.9 where a force of 5 kgf is applied to a piston of area 2 cm². This produces a pressure of 2.5 kgf cm^{-2} at every point within the fluid, which acts with equal force per unit area on the walls of the system.

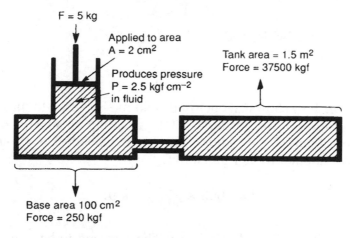

(a) Forces and pressure in closed tanks

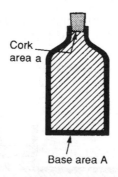

(b) Pressure in a bottle

Figure 1.9 *Pressure in an enclosed fluid*

Suppose the base of the left hand tank is 0.1×0.1 m to give a total area of 100cm². The total force acting on the base will be 250 kgf. If the top of the right hand tank is 1 m × 1.5 m, a surprisingly large upwards force of 37,500 kgf is developed. Note, the size of the connecting pipe has no effect. This principle explains why it is possible to shear the bottom off a bottle by applying a small force to the cork, as illustrated in Figure 1.9b.

The applied force develops a pressure, given by the expression:

$$P = \frac{f}{a}. \tag{1.8}$$

The force on the base is:

$$F = P \times A. \tag{1.9}$$

from which can be derived:

$$F = f \times \frac{A}{a}. \tag{1.10}$$

Expression 1.10 shows an enclosed fluid may be used to magnify a force. In Figure 1.10 a load of 2000 kg is sitting on a piston of area 500 cm² (about 12 cm radius). The smaller piston has an area of 2 cm². An applied force f given by:

$$f = 2000 \times \frac{2}{500} = 8 \text{ kgf.} \tag{1.11}$$

will cause the 2000 kg load to rise. There is said to be a *mechanical advantage* of 250.

Energy must, however, be conserved. To illustrate this, suppose the left hand piston moves down by 100 cm (one metre). Because

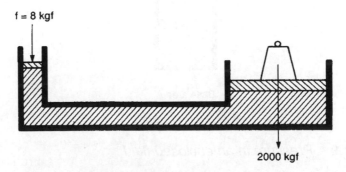

Figure 1.10   *Mechanical advantage*

we have assumed the fluid is incompressible, a volume of liquid 200 cm^2 is transferred from the left hand cylinder to the right hand cylinder, causing the load to rise by just 0.4 cm. So, although we have a force magnification of 250, we have a movement reduction of the same factor. Because work is given by the product of force and the distance moved, the force is magnified and the distance moved reduced by the same factor, giving conservation of energy. The action of Figure 1.10 is thus similar to the mechanical systems of Figure 1.11 which also exhibit mechanical advantage.

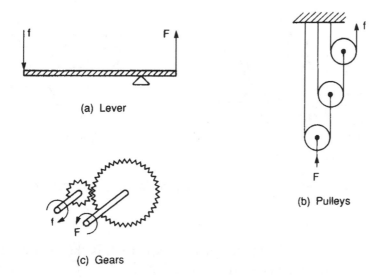

(a) Lever

(b) Pulleys

(c) Gears

Figure 1.11   *Examples of mechanical advantage where a small input force f produces a larger output force F*

The principle of Figure 1.10 is widely used where a large force is required with small movement. Typical examples are clamps, presses, hydraulic jacks and motor car brake and clutch operating mechanisms.

It should be noted that pressure in, say, a cylinder is determined solely by load and piston area in the steady state, and is not dependent on velocity of the piston once a constant speed has been achieved. Relationships between force, pressure, flow and speed are illustrated in Figure 1.12.

In Figure 1.12a, fluid is delivered to a cylinder at a rate of Q cm^3 s^{-1}. When the inlet valve is first opened, a pressure spike is observed as the load accelerates, but the pressure then settles back

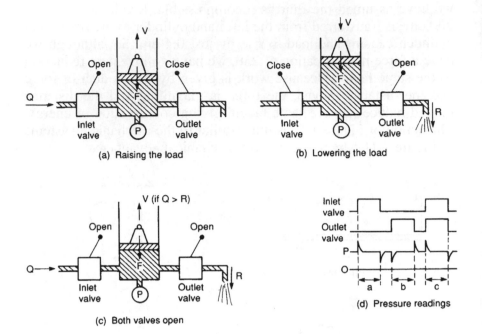

(a) Raising the load

(b) Lowering the load

(c) Both valves open

(d) Pressure readings

Figure 1.12   *The relationships between force, pressure, flow and speed*

to a steady value of $P = F/A$ kgf cm^{-2} where A is the area of the piston in cm^2 and F is measured in kgf. The load rises with a velocity $V = Q/A$ cm s^{-1} and velocity can obviously be controlled by adjusting flow rate Q.

In Figure 1.12b, the inlet valve has been closed, and the outlet valve opened allowing R cm^{-3} s^{-1} to flow out of the cylinder. There is again a pressure spike (negative this time) as the load accelerates downwards, but the pressure reverts to $P = F/A$ once the steady speed $V = R/A$ cm s^{-1} is achieved.

Finally, in Figure 1.12c both valves are open. The net flow is (Q–R) giving a cylinder velocity (Q–R)/A which can be positive (rising) or negative (falling) dependent on which flow is the largest. The steady state pressure, however, is unchanged at $P = F/A$.

Pressure measurement

Behaviour of a fluid can generally be deduced from measurements of flow or pressure. A flow transducer or transmitter has to be plumbed, in line, into a pipe, whereas pressure transmitters can be added non-intrusively as tappings to the side of a pipe. The basic fault-finding tool in both pneumatic or hydraulic systems is therefore a pressure gauge. Often this is a simple gauge which can be plugged into various parts of the system via a flexible connection.

These test pressure gauges invariably measure gauge pressure with the simple Bourdon pressure gauge shown in Figure 1.13. This consists of a flattened C shaped tube which is fixed at one end, shown in Figure 1.13a. When pressure is applied to the tube it tends to straighten, with the free end moving up and to the right. For low pressure ranges a spiral tube is used to increase the sensitivity.

This movement is converted to a circular pointer movement by a mechanical quadrant and pinion. If an electrical output signal is required for remote indication, the pointer can be replaced by a potentiometer, as shown in Figure 1.13b.

Hydraulic and pneumatic systems tend to exhibit large pressure spikes as loads accelerate or decelerate (a typical example being shown on Figure 1.12c.) These spikes can be irritating to the observer, can mislead, and in extreme cases could damage a pressure indicator. The response of a pressure sensor can be dampened by inclusion of a snubber restriction, as shown in Figure 1.13c.

Bourdon gauge-based transducers are generally robust but are low accuracy (typically ± 2%) devices. As the limit of visual resolution of a pointer position is no better than ± 2% anyway, ruggedness of these transducers makes them ideal for plant mounted monitoring.

Where more accurate pressure measurement is required, transducers based on the force balance principle of Figure 1.14 are generally used. This is essentially a differential pressure transducer, in which the low pressure inlet (LP) is left open to atmosphere and the high pressure (HP) inlet connects to the system. The signal given (HP-LP) is thus gauge pressure.

A pressure increase in the system deflects the pressure sensitive diaphragm to the left. This movement is detected by the displace-

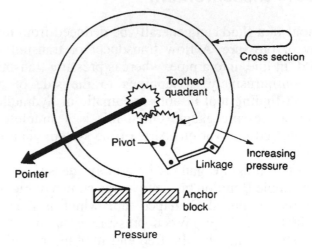

(a) Bourdon tube gauge construction

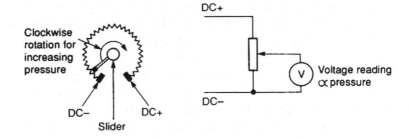

(b) Electrical signal from Bourdon gauge

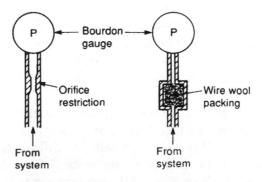

(c) Snubber restrictions

Figure 1.13 *The Bourdon pressure gauge*

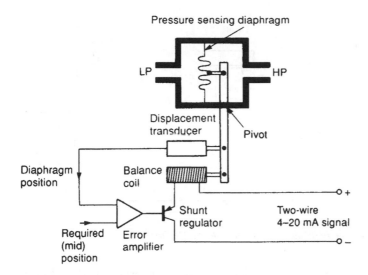

Figure 1.14 *Force balance pressure transducer*

ment transducer which, via a servo amplifier, leads to an increase in current in the balance coil.

Because the force from the balance coil always exactly balances the force arising from the pressure difference between LP and HP, current through the transducer is directly proportional to the differential pressure.

Remote indicating transducers are generally arranged with a remote power supply and the indicator and/or recorder connected into one line as Figure 1.15 to give a two-wire system. A signal range of 4 to 20 mA is commonly used, with the 4 mA zero level providing a current supply for the transducer's servo amplifier and also indicating circuit continuity (0 mA indicating a open circuit fault condition).

Fluid flow

Hydraulic and pneumatic systems are both concerned with the flow of a fluid (liquid or gas) down a pipe. Flow is a loose term that generally has three distinct meanings:

- *volumetric flow is* used to measure volume of fluid passing a point per unit of time. Where the fluid is a compressible gas, temperature and pressure must be specified or flow normalised to

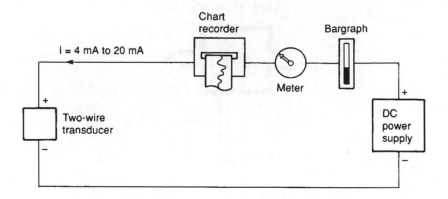

Figure 1.15 *Advantages of two-wire transducers*

some standard temperature and pressure (a topic discussed later). Volumetric flow is the most common measurement in process control
* *mass flow* measures the mass of fluid passing the point in unit time
* *velocity of flow* measures linear speed (in m s^{-1}, say) past the point of measurement. Flow velocity is of prime importance in the design of hydraulic and pneumatic systems.

Types of fluid flow are illustrated in Figure 1.16. At low flow velocities, the flow pattern is smooth and linear with low velocities at the pipe walls and the highest flow at the centre of the pipe. This is known as *laminar* or *streamline* flow.

As flow velocity increases, eddies start to form until at high flow velocities complete turbulence results as shown in Figure 1.16b. Flow velocity is now virtually uniform across the pipe.

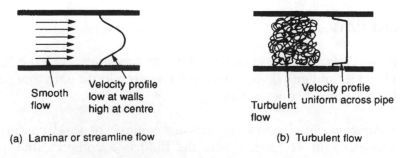

Figure 1.16 *Types of fluid flow*

The nature of the flow is determined by the *Reynolds number,* R_c, given by the expression:

$$R_c = \frac{vd\rho}{\eta}.$$ (1.12)

where v is flow velocity, d is pipe diameter, ρ the fluid density and η the viscosity. The Reynolds number is a ratio and hence dimensionless. If $R_c < 2000$, flow is laminar. If $R_c > 10^5$, flow is turbulent.

A turbulent flow is generally preferred for products in process control as it simplifies volumetric flow measurement (with differential pressure flowmeters – see later). Turbulent flow, however, increases energy loss through friction and may lead to premature wear. Cavitation (formation and collapse of vapour bubbles) occurs with turbulent liquid flow and may result in pitting on valve surfaces. Laminar flow is therefore specified for hydraulic and pneumatic systems. This results in a desired flow velocity of about 5 m s^{-2}.

Energy in a unit mass of fluid has three components:

- kinetic energy from its motion, given by $v^2/2$ where v is flow velocity
- potential energy from the height of the fluid
- energy arising from the pressure of the fluid, given by P/ρ where P is the pressure and ρ the density.

Fluid is passing along a pipe in Figure 1.17. Neglecting energy losses from friction, energies at points X, Y and Z will be equal. The flow velocity at point Y, however, is higher than at points X and Z

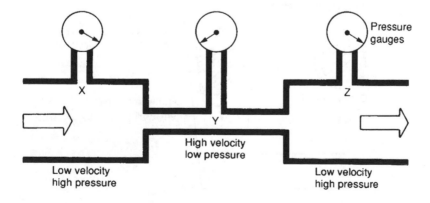

Figure 1.17 *Relationship between flow and pressure*

because of the smaller pipe diameter. Potential energy at each point is constant because the pipe is horizontal. so we can write:

$$\frac{v_x^2}{2} + \frac{P_x}{\rho} = \frac{v_y^2}{2} + \frac{P_y}{\rho} = \frac{v_z^2}{2} + \frac{P_z}{\rho} \qquad (1.13)$$

$$\underset{\text{at X}}{\text{Energy}} \quad \underset{\text{at Y}}{\text{Energy}} \quad \underset{\text{at Z}}{\text{Energy}}$$

We have implied an incompressible fluid by assuming the density, ρ, is constant throughout. Expression 1.13 becomes more complicated for a gas as different densities have to be used at each point.

The net result of the expression is fluid pressure falls as flow velocity rises. Note, though, that the pressure recovers as flow velocity falls again at point Z.

The simplest method of measuring flow (known as a variable area flowmeter) uses a float in a vertical tube arranged as Figure 1.18. The obstruction of the float causes a local increase in the fluid velocity which causes a differential pressure drop across the float, resulting in an upward force. The weight of the float obviously causes a downward force. The float therefore rises or falls depending on which force is the largest. The area around the float, however, increases the higher the float rises because of the tube taper. This increase in area decreases the pressure drop across the float and the upwards force. The float therefore settles at a vertical

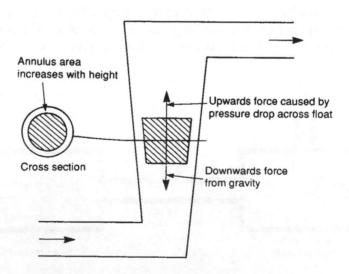

Figure 1.18 *Variable area flowmeter*

position where the weight of the float and the upwards force from the differential pressure exactly match. Flow rate can therefore be determined from the float position.

A remote indicating flowmeter can be constructed from a pipe mounted turbine, as shown in Figure 1.19. Fluid flow causes the propeller to rotate, its rotational speed being proportional to flow rate. Blade rotation is counted electronically by an external inductive proximity detector to give an electrical signal for remote indication of the flow rate.

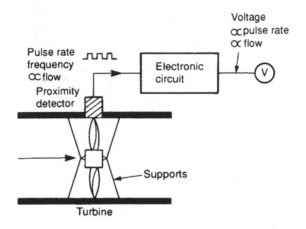

Figure 1.19 *Turbine flowmeter*

Finally, the classical method of measuring flow returns directly to expression 1.13 by locally increasing flow velocity with a deliberately introduced restriction as shown in Figure 1.20. Typical obstructions are an orifice plate or a venturi. These increase flow velocity, causing a pressure drop which can be measured to give a differential pressure related to the flow. Unfortunately, the differential pressure is proportional to the square of the flow rate, so a linearising square root extractor circuit is required to give a linear signal. Although differential pressure flow measurement is widely used to measure the flow rates of process material, the technique is not widely used in hydraulic and pneumatic systems.

It will be apparent that all flow measurement systems are intrusive to various degrees, and cannot be tapped in as easily as pressure measurement can. Fault finding in hydraulic and pneumatic systems is therefore generally based on pressure readings at strategic points.

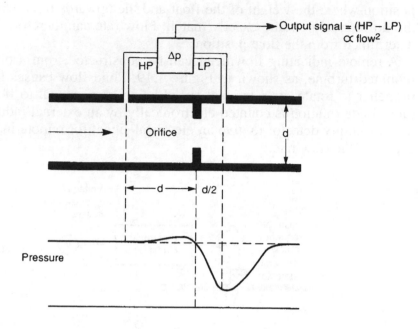

Figure 1.20   *Orifice plate flowmeter*

Temperature

Fluid behaviour is determined to some extent by its temperature. A later section discusses the relationship between pressure and temperature in a gas.

Temperature scales

A temperature scale is established by choosing two observable physical effects which are dependent upon temperature and assigning numerical values to them. The Fahrenheit and Celsius (previously known as Centigrade) scales use the freezing and boiling points of water as the two reference points:

	Fahrenheit	Celsius
Freezing point	32	0
Boiling point	212	100

From which:

$$F = \left(9 \times \frac{C}{5}\right) + 32. \qquad (1.14)$$

and:

$$C = (F - 32) \times \frac{5}{9}. \qquad (1.15)$$

The SI unit of temperature is the Kelvin. This defines the lowest theoretical temperature (called *absolute zero*) as 0 K, and the triple point of water (0.01°C) as 273.16 K. It should be noted that temperatures in Kelvin do not use the degree (°) symbol. These apparently odd numerical values make a temperature change of 1 K the same as 1°C, and:

$$K = °C + 273.1. \qquad (1.16)$$

The Celsius scale is most widely used in industry, but the Kelvin scale is important in determining the changes in gas pressure or volume with temperature.

Temperature measurement

There are four basic ways of measuring temperature based on temperature-dependent physical properties.

Expansion of a substance with temperature can be used to produce a change in volume, length or pressure. This is probably the most common type of temperature measurement in the form of mercury or alcohol-in-glass thermometers. A variation is the bimetallic strip shown in Figure 1.21. where two dissimilar metals have different coefficients of expansion which cause the strip to

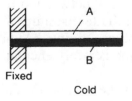

Fixed

Cold

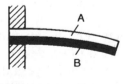

Hot, metal A expands
more than metal B

Figure 1.21 *Bimetallic strip*

bend according to the temperature. This technique is the basis of most on/off thermostats used for temperature control or alarm annunciation. A bimetallic spiral can be used to construct an indicating thermometer.

Electrical resistance changes with temperature. A platinum wire with resistance 100 ohms at 0°C will have a resistance of 138.5 ohms at 100°C. Temperature sensors based on this principle are known as RTDs (for resistance temperature detector) or PT100 sensors (from PT, for platinum, and 100 for 100 ohms at 0°C). Semiconductor devices called thermistors have more dramatic changes, the characteristics of a typical device being shown in Figure 1.22. The response, however, is non-linear which makes thermistors more suitable for alarm/control application than temperature indication.

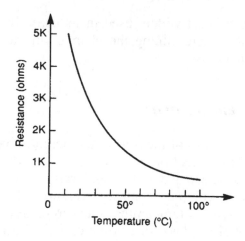

Figure 1.22   *Typical resistance temperature curve for NTC thermistor*

Thermocouples, the principle of which is shown in Figure 1.23, use the small difference in contact potentials between different metals to give a voltage which depends on the temperature difference between the measurement and reference points. Although widely used in process control, the technique is rarely encountered in pneumatic and hydraulic systems.

The final method, called pyrometry, uses the change in radiated energy with temperature. As this has a minimum temperature measurement of about 400°C, it is totally unsuitable for the systems we shall be discussing.

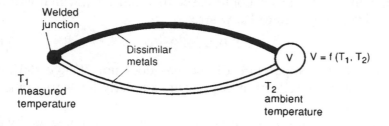

Figure 1.23 *The thermocouple*

Gas laws

For all practical purposes, liquids used in hydraulic systems can be considered incompressible and insensitive to changes in temperature (provided the temperature remains within some quite broad limits). The gas in a pneumatic system is very sensitive to changes in pressure and temperature, and its behaviour is determined by the gas laws described below.

In the following expressions it is important to note that pressures are given in absolute, not gauge, terms and temperatures are given in absolute degrees Kelvin, not in degrees Celsius. If we discuss, say, a litre of air at atmospheric pressure and 20°C being compressed to three atmospheres gauge pressure, its original pressure was one atmosphere, its original temperature was 293 K and its final pressure is four atmospheres absolute.

Pressure and volume are related by Boyle's law. In Figure 1.24 we have a volume of gas V_1 at pressure P_1 (in absolute units,

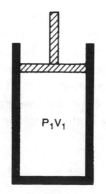
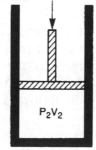

Figure 1.24 *Boyle's law*

remember). This gas is compressed to volume V_2, which will result in a rise of pressure to P_2, where:

$$P_1V_1 = P_2V_2. \qquad (1.17)$$

provided the temperature of the gas does not change during the compression. A reduction of pressure similarly leads to an increase in volume.

In practice, compression of a gas is always accompanied by a rise in temperature (as is commonly noticed when pumping up a bicycle tyre) and a reduction in pressure produces a temperature fall (the principle of refrigeration). For expression 1.17 to apply, the gas must be allowed to return to its original temperature.

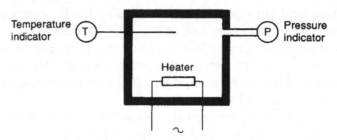

Figure 1.25   *Relationship between temperature and pressure*

In Figure 1.25, on the other hand, the temperature of a fixed volume of gas is controlled by a heater. A rise in temperature from T_1 to T_2 results in an increase in pressure from P_1 to P_2, where:

$$\frac{P_1}{T_1} = \frac{P_2}{T_2}. \qquad (1.18)$$

Again it should be remembered pressure and temperature are in absolute terms. Although expression 1.18 gives the change in pressure resulting from a change in temperature, it also applies to changes of temperature resulting from a change in pressure provided no heat is lost from the system. In a pneumatic air compressor, the temperature of the outgoing compressed air is considerably elevated by the increase in pressure, resulting in the need for the compressor to be followed by an air cooler.

Expressions 1.17 and 1.18 are combined to give the general gas law:

$$\frac{P_1 V_1}{T_1} = \frac{P_2 V_2}{T_2} \qquad (1.19)$$

where P_1, V_1, T_1 are initial conditions and P_2, V_2, T_2 are final conditions. As before, expression 1.19 assumes no heat is lost to, or gained from, the environment.

2

Hydraulic pumps and pressure regulation

A hydraulic pump (Figure 2.1) takes oil from a tank and delivers it to the rest of the hydraulic circuit. In doing so it raises oil pressure to the required level. The operation of such a pump is illustrated in Figure 2.1a. On hydraulic circuit diagrams a pump is represented by the symbol of Figure 2.1b, with the arrowhead showing the direction of flow.

Hydraulic pumps are generally driven at constant speed by a three phase AC induction motor rotating at 1500 rpm in the UK (with a 50 Hz supply) and at 1200 or 1800 rpm in the USA (with a 60 Hz supply). Often pump and motor are supplied as one combined unit. As an AC motor requires some form of starter, the complete arrangement illustrated in Figure 2.1c is needed.

There are two types of pump (for fluids) or compressor (for gases) illustrated in Figure 2.2. Typical of the first type is the centrifugal pump of Figure 2.2a. Fluid is drawn into the axis of the pump, and flung out to the periphery by centrifugal force. Flow of fluid into the load maintains pressure at the pump exit. Should the pump stop, however, there is a direct route from outlet back to inlet and the pressure rapidly decays away. Fluid leakage will also occur past the vanes, so pump delivery will vary according to outlet pressure. Devices such as that shown in Figure 2.2a are known as hydrodynamic pumps, and are primarily used to shift fluid from one location to another at relatively low pressures. Water pumps are a typical application.

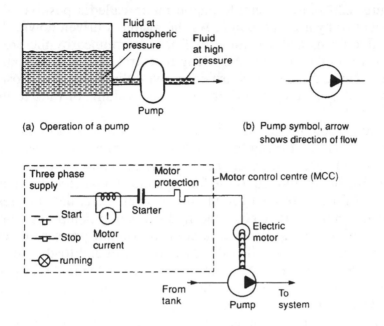

(a) Operation of a pump

(b) Pump symbol, arrow shows direction of flow

(c) Pump associated components

Figure 2.1   *The hydraulic pump*

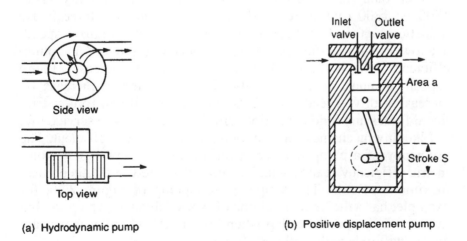

(a) Hydrodynamic pump

(b) Positive displacement pump

Figure 2.2   *Types of hydraulic pump*

Figure 2.2b shows a simple piston pump called a positive displacement or hydrostatic pump. As the piston is driven down, the inlet valve opens and a volume of fluid (determined by the cross section area of the piston and the length of stroke) is drawn into the cylinder. Next, the piston is driven up with the inlet valve closed and the outlet valve open, driving the same volume of fluid to the pump outlet.

Should the pump stop, one of the two valves will always be closed, so there is no route for fluid to leak back. Exit pressure is therefore maintained (assuming there are no downstream return routes).

More important, though, is the fact that the pump delivers a fixed volume of fluid from inlet to outlet each cycle regardless of pressure at the outlet port. Unlike the hydrodynamic pump described earlier, a piston pump has no inherent maximum pressure determined by pump leakage: if it drives into a dead end load with no return route (as can easily occur in an inactive hydraulic system with all valves closed) the pressure rises continuously with each pump stroke until either piping or the pump itself fails.

Hydraulic pumps are invariably hydrostatic and, consequently, require some method of controlling system pressure to avoid catastrophic pipe or pump failure. This topic is discussed further in a later section.

A hydraulic pump is specified by the flow rate it delivers (usually given in litres min^{-1} or gallons min^{-1}) and the maximum pressure the pump can withstand. These are normally called the pump capacity (or delivery rate) and the pressure rating.

Pump data sheets specify required drive speed (usually 1200, 1500 or 1800 rpm corresponding to the speed of a three phase induction motor). Pump capacity is directly related to drive speed; at a lower than specified speed, pump capacity is reduced and pump efficiency falls as fluid leakage (called slippage) increases. Pump capacity cannot, on the other hand, be expected to increase by increasing drive speed, as effects such as centrifugal forces, frictional forces and fluid cavitation will drastically reduce service life.

Like any mechanical device, pumps are not 100% efficient. The efficiency of a pump may be specified in two ways. First, volumetric efficiency relates actual volume delivered to the theoretical maximum volume. The simple piston pump of Figure 2.2b, for example, has a theoretical volume of $A \times s$ delivered per stroke, but in practice the small overlap when both inlet and outlet valves are closed will reduce the volume slightly.

Second, efficiency may be specified in terms of output hydraulic power and input mechanical (at the drive shaft) or electrical (at the motor terminals) power.

Typical efficiencies for pumps range from around 90% (for cheap gear pumps) to about 98% for high quality piston pumps. An allowance for pump efficiency needs to be made when specifying pump capacity or choosing a suitable drive motor.

The motor power required to drive a pump is determined by the pump capacity and working pressure. From expression 1.6:

$$\text{Power} = \frac{\text{work}}{\text{time}}$$

$$= \frac{\text{force} \times \text{distance}}{\text{time}}.$$

In Figure 2.3, a pump forces fluid along a pipe of area A against a pressure P, moving fluid a distance d in time T. The force is PA, which, when substituted into expression 2.1, gives:

$$\text{Power} = \frac{P \times A \times d}{T}$$

but A × d/T is flow rate, hence:

$$\text{Power} = \text{pressure} \times \text{flow rate}. \tag{2.2}$$

Unfortunately, expression 2.2 is specified in impractical SI units (pressure in pascal, time in seconds, flow in cubic metres). We may adapt the expression to use more practical units (pressure in bar, flow rate in litres min^{-1}) with the expression:

$$\text{Power} = \frac{\text{pressure} \times \text{flow rate}}{600} \text{ Kw.} \tag{2.3}$$

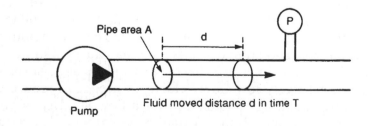

Figure 2.3 *Derivation of pump power*

For Imperial systems (pressure in psig, flow rate in gallons min^{-1}), the expression becomes:

$$\text{Power} = \frac{\text{pressure} \times \text{flow rate}}{1915} \text{ Kw.} \tag{2.4}$$

For fully Imperial systems, motor power in horsepower can be found from:

$$\text{Horsepower} = 0.75 \times \text{power in Kw.} \tag{2.5}$$

Hydraulic pumps such as that in Figure 2.1 do not require priming because fluid flows, by gravity, into the pump inlet port. Not surprisingly this is called a self-priming pump. Care must be taken with this arrangement to avoid sediment from the tank being drawn into the pump.

The pump in Figure 2.4 is above the fluid in the tank. The pump creates a negative (less than atmospheric) pressure at its inlet port causing fluid to be pushed up the inlet pipe by atmospheric pressure. This action creates a fluid lift which is, generally, incorrectly described as arising from pump suction. In reality fluid is *pushed* into the pump.

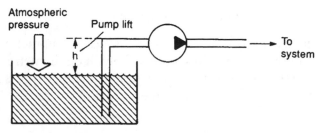

Figure 2.4   *Pump lift*

Maximum pump lift is determined by atmospheric pressure and is given by expressions 1.3 and 1.4. In theory a lift of about 8 m is feasible but, in practice, would be accompanied by undesirable side effects such as cavitation (formation and destructive collapse of bubbles from partial vaporisation of fluid). The lift should be as small as possible and around 1 m is a normal practical limit.

Fluid flow in the inlet line always takes place at negative pressure, and a relatively low flow velocity is needed to reduce these side effects. The design should aim for a flow velocity of around 1 m s^{-1}. Examination of any hydraulic system will always reveal pump inlet pipes of much larger diameters than outlet pipes.

Pressure regulation

Figure 2.5a shows the by now familiar system where a load is raised or lowered by a hydraulic cylinder. With valve V_1 open, fluid flows from the pump to the cylinder, with both pressure gauges P_1 and P_2 indicating a pressure of F/A. With valves V_1 closed and V_2 open, the load falls with fluid being returned to the tank. With the load falling, gauge P_2 will still show a pressure of F/A, but at P_1 the pump is dead-ended leading to a continual increase in pressure as the pump delivers fluid into the pipe.

Obviously some method is needed to keep P_1 at a safe level. To achieve this, pressure regulating valve V_3 has been included. This is normally closed (no connection between P and T) while the pressure is below some preset level (called the cracking pressure). Once the cracking pressure is reached valve V_3 starts to open, bleeding fluid back to the tank. As the pressure increases, valve V_3 opens more until, at a pressure called the full flow pressure, the valve is

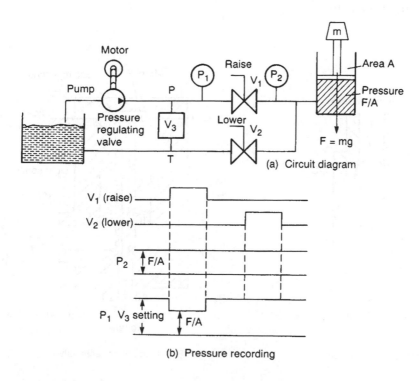

(a) Circuit diagram

(b) Pressure recording

Figure 2.5 *Action of pressure regulation*

fully open. With valve V_1 closed, all fluid from the pump returns to
the tank via the pressure regulating valve, and P_1 settles somewhere
between the cracking and full flow pressures.

Cracking pressure of a relief valve *must* be higher than a system's
working pressure, leading to a fall in system pressure as valve V_1
opens and external work is performed. Valve positions and conse-
quent pressure readings are shown in Figure 2.5b.

The simplest form of pressure regulation valve is the ball
and spring arrangement of Figure 2.6a. System pressure in
the pipe exerts a force of $P \times a$ on the ball. When the force is larger
than the spring compressive force the valve will crack open,
bypassing fluid back to the tank. The higher the pipe pressure, the
more the valve opens. Cracking pressure is set by the spring compression and in practical valves this can be adjusted to suit the
application.

The difference between cracking and full flow pressure is called
the pressure override. The steady (non-working) system pressure
will lie somewhere within the pressure override, with the actual
value determined by pipe sizes and characteristics of the pressure
regulating valve itself.

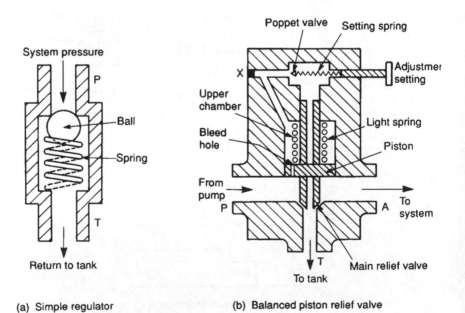

(a) Simple regulator   (b) Balanced piston relief valve

Figure 2.6   *Pressure regulation*

If the quiescent pressure is required to be precisely defined, a small pressure override is needed. This pressure override is related to spring tension in a simple relief valve. When a small, or precisely defined, override is required, a balanced piston relief valve (shown in Figure 2.6b) is used.

The piston in this valve is free moving, but is normally held in the lowered position by a light spring, blocking flow to the tank. Fluid is permitted to pass to the upper chamber through a small hole in the piston. The upper chamber is sealed by an adjustable spring-loaded poppet. In the low pressure state, there is no flow past the poppet, so pressure on both sides of the piston are equal and spring pressure keeps the valve closed.

When fluid pressure rises, the poppet cracks and a small flow of fluid passes from the upper chamber to the tank via the hole in the piston centre. This fluid is replenished by fluid flowing through the hole in the piston. With fluid flow there is now a pressure differential across the piston, which is acting only against a light spring. The whole piston lifts, releasing fluid around the valve stem until a balance condition is reached. Because of the light restoring spring a very small override is achieved.

The balanced piston relief valve can also be used as an unloading valve. Plug X is a vent connection and, if removed, fluid flows from the main line through the piston. As before, this causes the piston to rise and flow to be dumped to the tank. Controlled loading/unloading can be achieved by the use of a finite position valve connected to the vent connection.

When no useful work is being performed, *all* fluid from the pump is pressurised to a high pressure then dumped back to the tank (at atmospheric pressure) through the pressure regulating valve. This requires motor power defined earlier by expression 2.3 and 2.4, and represents a substantial waste of power. Less obviously, energy put into the fluid is converted to heat leading to a rise in fluid temperature. Surprisingly, motor power will be higher when no work is being done because cracking pressure is higher than working pressure.

This waste of energy is expensive, and can lead to the need for heat exchangers to be built into the tank to remove the excess heat. A much more economic arrangement uses loading/unloading valves, a topic discussed further in a later section.

Pump types

There are essentially three different types of positive displacement pump used in hydraulic systems.

Gear pumps

The simplest and most robust positive displacement pump, having just two moving parts, is the gear pump. Its parts are non-reciprocating, move at constant speed and experience a uniform force. Internal construction, shown in Figure 2.7, consists of just two close meshing gear wheels which rotate as shown. The direction of rotation of the gears should be carefully noted; it is the *opposite* of that intuitively expected by most people.

As the teeth come out of mesh at the centre, a partial vacuum is formed which draws fluid into the inlet chamber. Fluid is trapped between the outer teeth and the pump housing, causing a continual transfer of fluid from inlet chamber to outlet chamber where it is discharged to the system.

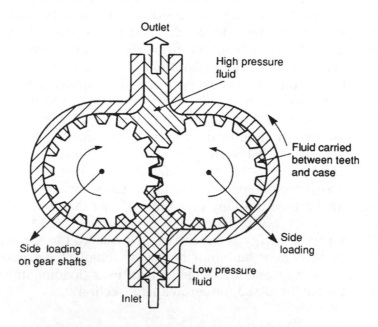

Figure 2.7   *Gear pump*

Pump displacement is determined by: volume of fluid between each pair of teeth; number of teeth; and speed of rotation. Note the pump merely delivers a fixed volume of fluid from inlet port to outlet port for each rotation; outlet port pressure is determined solely by design of the rest of the system.

Performance of any pump is limited by leakage and the ability of the pump to withstand the pressure differential between inlet and outlet ports. The gear pump obviously requires closely meshing gears, minimum clearance between teeth and housing, and also between the gear face and side plates. Often the side plates of a pump are designed as deliberately replaceable wear plates. Wear in a gear pump is primarily caused by dirt particles in the hydraulic fluid, so cleanliness and filtration are particularly important.

The pressure differential causes large side loads to be applied to the gear shafts at 45° to the centre line as shown. Typically, gear pumps are used at pressures up to about 150 bar and capacities of around 150 gpm (6751 min^{-1}). Volumetric efficiency of gear pumps at 90% is lowest of the three pump types.

There are some variations of the basic gear pump. In Figure 2.8, gears have been replaced by lobes giving a pump called, not surprisingly, a lobe pump.

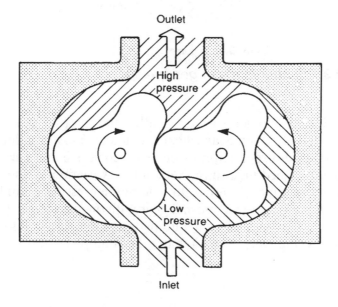

Figure 2.8 *The lobe pump*

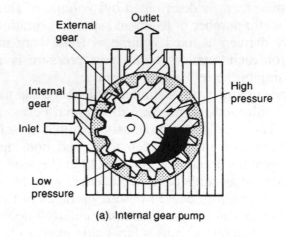

(a) Internal gear pump

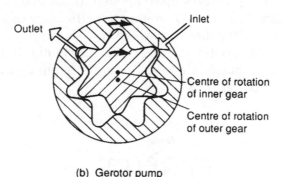

(b) Gerotor pump

Figure 2.9   *Further forms of gear pump*

Figure 2.9a is another variation called the internal gear pump, where an external driven gear wheel is connected to a smaller internal gear, with fluid separation as gears disengage being performed by a crescent-shaped moulding. Yet another variation on the theme is the gerotor pump of Figure 2.9b, where the crescent moulding is dispensed with by using an internal gear with one less tooth than the outer gear wheel. Internal gear pumps operate at lower capacities and pressures (typically 70 bar) than other pump types.

Vane pumps

The major source of leakage in a gear pump arises from the small gaps between teeth, and also between teeth and pump housing. The vane pump reduces this leakage by using spring (or hydraulic) loaded vanes slotted into a driven rotor, as illustrated in the two examples of Figure 2.10.

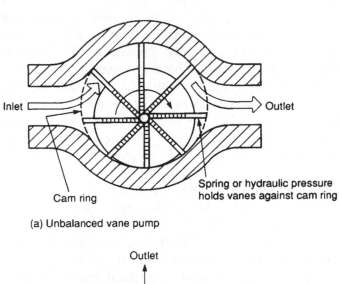

(a) Unbalanced vane pump

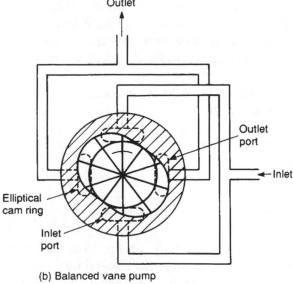

(b) Balanced vane pump

Figure 2.10 *Vane pumps*

In the pump shown in Figure 2.10a, the rotor is offset within the housing, and the vanes constrained by a cam ring as they cross inlet and outlet ports. Because the vane tips are held against the housing there is little leakage and the vanes compensate to a large degree for wear at vane tips or in the housing itself. There is still, however, leakage between rotor faces and body sides. Pump capacity is determined by vane throw, vane cross sectional area and speed of rotation.

The difference in pressure between outlet and inlet ports creates a severe load on the vanes and a large side load on the rotor shaft which can lead to bearing failure. The pump in Figure 2.10a is consequently known as an unbalanced vane pump. Figure 2.10b shows a balanced vane pump. This features an elliptical cam ring together with two inlet and two outlet ports. Pressure loading still occurs in the vanes but the two identical pump halves create equal but opposite forces on the rotor, leading to zero net force in the shaft and bearings. Balanced vane pumps have much improved service lives over simpler unbalanced vane pumps.

Capacity and pressure ratings of a vane pump are generally lower than gear pumps, but reduced leakage gives an improved volumetric efficiency of around 95%.

In an ideal world, the capacity of a pump should be matched exactly to load requirements. Expression 2.2 showed that input power is proportional to system pressure and volumetric flow rate. A pump with too large a capacity wastes energy (leading to a rise in fluid temperature) as excess fluid passes through the pressure relief valve.

Pumps are generally sold with certain fixed capacities and the user has to choose the next largest size. Figure 2.11 shows a vane pump with adjustable capacity, set by the positional relationship between rotor and inner casing, with the inner casing position set by an external screw.

Piston pumps

A piston pump is superficially similar to a motor car engine, and a simple single cylinder arrangement was shown earlier in Figure 2.2b. Such a simple pump, however, delivering a single pulse of fluid per revolution, generates unacceptably large pressure pulses into the system. Practical piston pumps therefore employ multiple cylinders

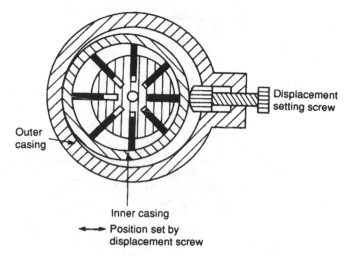

Figure 2.11 *Variable displacement vane pump*

and pistons to smooth out fluid delivery, and much ingenuity goes into designing multicylinder pumps which are surprisingly compact.

Figure 2.12 shows one form of radial piston pump. The pump consists of several hollow pistons inside a stationary cylinder block. Each piston has spring-loaded inlet and outlet valves. As the inner cam rotates, fluid is transferred relatively smoothly from inlet port to the outlet port.

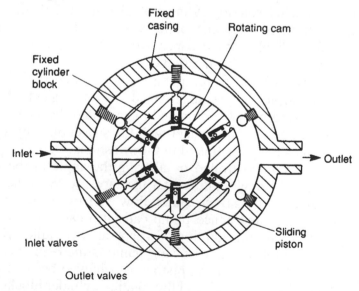

Figure 2.12 *Radial piston pump*

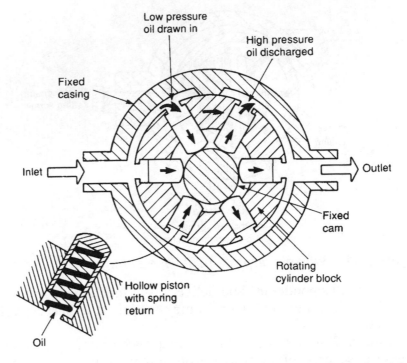

Figure 2.13   *Piston pump with stationary cam and rotating block*

The pump of Figure 2.13 uses the same principle, but employs a stationary cam and a rotating cylinder block. This arrangement does not require multiple inlet and outlet valves and is consequently simpler, more reliable, and cheaper. Not surprisingly most radial piston pumps have this construction.

An alternative form of piston pump is the axial design of Figure 2.14, where multiple pistons are arranged in a rotating cylinder. The pistons are stroked by a fixed angled plate called the swash plate. Each piston can be kept in contact with the swash plate by springs or by a rotating shoe plate linked to the swash plate.

Pump capacity is controlled by altering the angle of the swash plate; the larger the angle, the greater the capacity. With the swash plate vertical capacity is zero, and flow can even be reversed. Swash plate angle (and hence pump capacity) can easily be controlled remotely with the addition of a separate hydraulic cylinder.

An alternative form of axial piston pump is the bent axis pump of Figure 2.15. Stroking of the pistons is achieved because of the angle between the drive shaft and the rotating cylinder block. Pump capacity can be adjusted by altering the drive shaft angle.

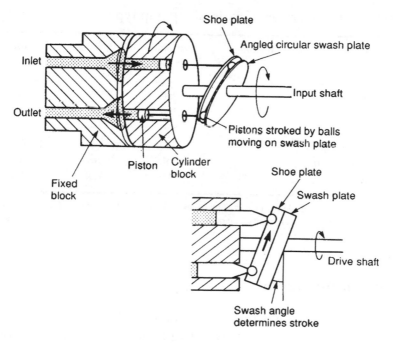

Figure 2.14   *Axial pump with swash plate*

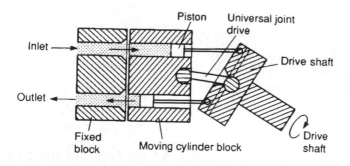

Figure 2.15   *Bent axis pump*

Piston pumps have very high volumetric efficiency (over 98%) and can be used at the highest hydraulic pressures. Being more complex than vane and gear pumps, they are correspondingly more expensive. Table 2.1 gives a comparison of the various types of pump.

Table 2.1   Comparison of hydraulic pump types

Type	Maximum pressure (bar)	Maximum flow (l/min)	Variable displacement	Positive displacement
Centrifugal	20	3000	No	No
Gear	175	300	No	Yes
Vane	175	500	Yes	Yes
Axial piston (port-plate)	300	500	Yes	Yes
Axial piston (valved)	700	650	Yes	Yes
In-line piston	1000	100	Yes	Yes

Specialist pumps are available for pressures up to about 7000 bar at low flows. The delivery from centrifugal and gear pumps can be made variable by changing the speed of the pump motor with a variable frequency (VF) drive.

Combination pumps

Many hydraulic applications are similar to Figure 2.16, where a workpiece is held in place by a hydraulic ram. There are essentially two distinct requirements for this operation. As the cylinder extends or retracts a large volume of fluid is required at a low pressure (sufficient just to overcome friction). As the workpiece is gripped, the requirement changes to a high pressure but minimal fluid volume.

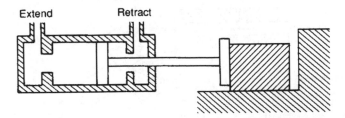

Figure 2.16 *A clamping cylinder. A large flow, but low pressure, is needed during extension and retraction, but zero flow and high pressure are needed during clamping*

This type of operation is usually performed with two separate pumps driven by a common electric motor as shown in Figure 2.17. Pump P_1 is a high pressure low volume pump, while pump P_2 is a high volume low pressure pump. Associated with these are two relief valves RV_1 and RV_2 and a one-way check (or non-return)

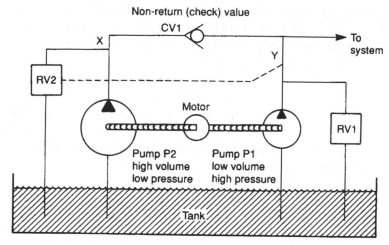

Figure 2.17   *Combination pump*

valve which allows flow from left to right, but blocks flow in the reverse direction.

A normal (high pressure) relief valve is used at position RV_1 but relief valve RV_2 is operated not by the pressure at point X, but remotely by the pressure at point Y. This could be achieved with the balanced piston valve of Figure 2.6. In low pressure mode both relief valves are closed and both pumps P_1 and P_2 deliver fluid to the load, the majority coming from pump P_2 because of its higher capacity.

When the workpiece is gripped, the pressure at Y rises, and relief valve RV_2 opens causing all the fluid from pump P_2 to return straight to the tank and the pressure at X to fall to a low value. Check valve CV_1 stops fluid from pump P_1 passing back to the tank via relief valve RV_2, consequently pressure at Y rises to the level set by relief valve RV_1.

This arrangement saves energy as the large volume of fluid from pump P_2 is returned to the tank at a very low pressure, and only a small volume of fluid from pump P_1 is returned at a high pressure. Pump assemblies similar to that shown in Figure 2.17 are called combination pumps and are manufactured as complete units with motor, pumps, relief and check valves prefitted.

Loading valves

Expression 2.2 shows that allowing excess fluid from a pump to return to the tank by a pressure relief valve is wasteful of energy and can lead to a rapid rise in temperature of the fluid as the wasted energy is converted to heat. It is normally undesirable to start and stop the pump to match load requirements, as this causes shock loads to pump, motor and couplings.

In Figure 2.18, valve V_1 is a normal pressure relief valve regulating pressure and returning excess fluid to the tank as described in earlier sections. The additional valve V_2 is opened or closed by an external electrical or hydraulic signal. With valve V_2 open, all the pump output flow is returned to the tank at low pressure with minimal energy cost.

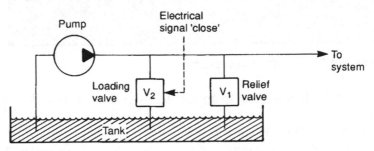

Figure 2.18   *Loading valve*

When fluid is required in the system the control signal closes valve V_2, pressure rises to the setting of valve V_1, and the system performs as normal. Valve V_2 is called a pump loading or a pump unloading valve according to the interpretation of the control signal sense.

Filters

Dirt in a hydraulic system causes sticking valves, failure of seals and premature wear. Even particles of dirt as small as 20 μ can cause damage, (1 micron is one millionth of a metre; the naked eye is just able to resolve 40 μ). Filters are used to prevent dirt entering the vulnerable parts of the system, and are generally specified in microns or meshes per linear inch (sieve number).

Inlet lines are usually fitted with strainers inside the tank, but these are coarse wire mesh elements only suitable for removing relatively large metal particles and similar contaminants Separate filters are needed to remove finer particles and can be installed in three places as shown in Figures 2.19a to c.

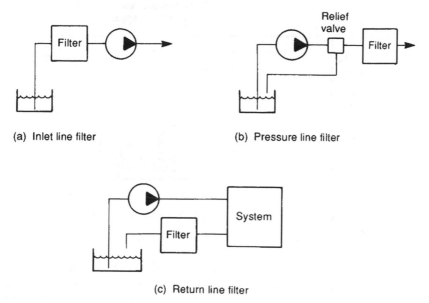

(a) Inlet line filter   (b) Pressure line filter

(c) Return line filter

Figure 2.19   *Filter positions*

Inlet line filters protect the pump, but must be designed to give a low pressure drop or the pump will not be able to raise fluid from the tank. Low pressure drop implies a coarse filter or a large physical size.

Pressure line filters placed after the pump protect valves and actuators and can be finer and smaller. They must, however, be able to withstand full system operating pressure. Most systems use pressure line filtering.

Return line filters may have a relatively high pressure drop and can, consequently, be very fine. They serve to protect pumps by limiting size of particles returned to the tank. These filters only have to withstand a low pressure. Filters can also be classified as full or proportional flow. In Figure 2.20a, all flow passes through the filter. This is obviously efficient in terms of filtration, but incurs a large pressure drop. This pressure drop increases as the filter becomes polluted, so a full flow filter usually incorporates a relief valve which cracks when the filter becomes unacceptably blocked. This is purely a safety feature, though, and the filter should, of course, have been changed before this state was reached as dirty unfiltered fluid would be passing round the system.

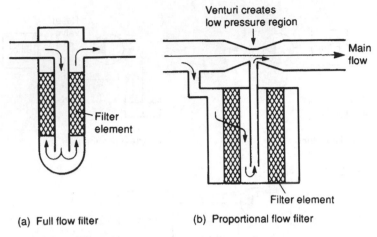

(a) Full flow filter   (b) Proportional flow filter

Figure 2.20   *Filter types*

In Figure 2.20b, the main flow passes through a venturi, creating a localised low pressure area. The pressure differential across the filter element draws a proportion of the fluid through the filter. This design is accordingly known as a proportional flow filter, as only a proportion of the main flow is filtered. It is characterized by a low pressure drop, and does not need the protection of a pressure relief valve.

Pressure drop across the filter element is an accurate indication of its cleanliness, and many filters incorporate a differential pressure meter calibrated with a green (clear), amber (warning), red (change overdue) indicator. Such types are called indicating filters.

Filtration material used in a filler may be mechanical or absorbent. Mechanical filters are relatively coarse, and utilise fine wire mesh or a disc/screen arrangement as shown in the edge type filter of Figure 2.21. Absorbent filters are based on porous materials such as paper, cotton or cellulose. Filtration size in an absorbent filter can be very small as filtration is done by pores in the material. Mechanical filters can usually be removed, cleaned and re-fitted, whereas absorbent filters are usually replaceable items.

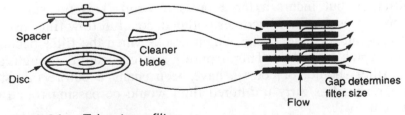

Figure 2.21   *Edge type filter*

In many systems where the main use is the application of pressure the actual draw from the tank is very small reducing the effectiveness of pressure and return line filters. Here a separate circulating pump may be used as shown on Figure 2.22 to filter and cool the oil. The running of this pump is normally a pre-condition for starting the main pumps. The circulation pump should be sized to handle the complete tank volume every 10 to 15 minutes.

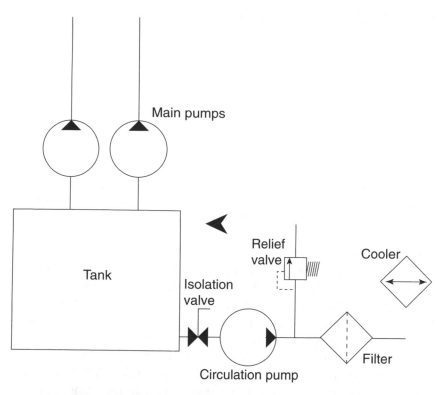

Figure 2.22 *A circulation pump used to filter and clean the fluid when the draw from the main pumps is small*

Note the pressure relief valve – this is included to provide a route back to tank if the filter or cooler is totally blocked. In a real life system additional hand isolation and non return valves would be fitted to permit changing the filter or cooler with the system running. Limit switches and pressure switches would also be included to signal to the control system that the hand isolation valves are open and the filter is clean.

3

Air compressors, air treatment and pressure regulation

The vast majority of pneumatic systems use compressed atmospheric air as the operating medium (a small number of systems use nitrogen obtained commercially from liquid gas suppliers). Unlike hydraulic systems, a pneumatic system is 'open'; the fluid is obtained free, used and then vented back to atmosphere.

Pneumatic systems use a compressible gas; hydraulic systems an incompressible liquid, and this leads to some significant differences. The pressure of a liquid may be raised to a high level almost instantaneously, whereas pressure rise in a gas can be distinctly leisurely. In Figure 3.1a, a reservoir of volume two cubic metres is connected to a compressor which delivers three cubic metres of air (measured at atmospheric pressure) per minute. Using Boyle's law (expression 1.17) the pressure rise shown in Figure 3.1b can be found.

Pressure in a hydraulic system can be quickly and easily controlled by devices such as unloading and pressure regulating valves. Fluid is thus stored at atmospheric pressure and compressed to the required pressure as needed. The slow response of an air compressor, however, precludes such an approach in a pneumatic system and necessitates storage of compressed air at the required pressure in a receiver vessel. The volume of this vessel is chosen so there are minimal deviations in pressure arising from flow changes in loads and the compressor is then employed to replace the air used, averaged over an extended period of time (e.g. a few minutes).

Deviations in air pressure are smaller, and compressor control is easier if a large receiver feeds many loads. A large number of loads

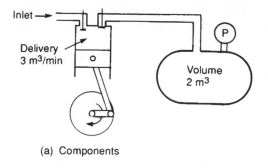

(a) Components

t (min)	Volume (at NTP)	P Abs	P gauge
0	2	1	0
1	5	2.5	1.5
2	8	4	3
3	11	5.5	4.5

(b) Response

Figure 3.1 *Compressibility of a gas*

statistically results in a more even flow of air from the receiver, also helping to maintain a steady pressure. On many sites, therefore, compressed air is produced as a central service which is distributed around the site in a similar manner to electricity, gas and water.

Behaviour of a gas subjected to changes in pressure, volume and temperature is governed by the general gas equation given earlier as expression 1.19 and reproduced here:

$$\frac{P_1 V_1}{T_1} = \frac{P_2 V_2}{T_2}.$$

(3.1)

where pressures are given in absolute terms and temperatures are measured in degrees Kelvin.

A compressor increases air pressure by reducing its volume, and expression 3.1 predicts a resultant rise in temperature. A pneumatic system must therefore incorporate some method of removing this excess heat. For small systems, simple fins on the compressor (similar in construction to an air-cooled internal combustion engine) will suffice. For larger systems, a separate cooler (usually employing water as the heat-removing medium) is needed.

Atmospheric air contains water vapour, the actual amount varying from day to day according to humidity. The maximum amount of water vapour held in a given volume of air is determined by temperature, and any excess condenses out as liquid droplets (commonly experienced as condensation on cold windows). A similar effect occurs as compressed air is cooled, and if left the resultant water droplets would cause valves to jam and corrosion to form in pipes. An aftercooler must therefore be followed by a water separator. Often aftercoolers and separators are called, collectively, primary air treatment units.

Dry cool air is stored in the receiver, with a pressure switch used to start and stop the compressor motor, maintaining the required pressure.

Ideally, air in a system has a light oil mist to reduce chances of corrosion and to lubricate moving parts in valves, cylinders and so on. This oil mist cannot be added before the receiver as the mist would form oil droplets in the receiver's relatively still air, so the exit air from the receiver passes through a unit which provides the lubricating mist along with further filtration and water removal. This process is commonly called secondary air treatment.

Often, air in the receiver is held at a slightly higher pressure than needed to allow for pressure drops in the pipe lines. A local pressure regulation unit is then employed with the secondary air treatment close to the device using air. Composite devices called service units comprising water separation, lubricator and pressure regulation are available for direct line monitoring close to the valves and actuators of a pneumatic system.

Figure 3.2 thus represents the components used in the production of a reliable source of compressed air.

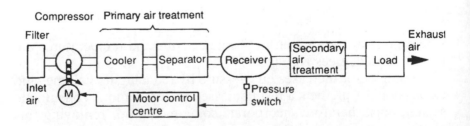

Figure 3.2   *Component parts of a pneumatic system*

Compressor types

Like hydraulic pumps, air compressors can be split into positive displacement devices (where a fixed volume of air is delivered on each rotation of the compressor shaft) and dynamic devices such as centrifugal or axial blowers. The vast majority of air compressors are of the positive displacement type.

A compressor is selected by the pressure it is required to work at and the volume of gas it is required to deliver. As explained in the previous section, pressure in the receiver is generally higher than that required at the operating position, with local pressure regulation being used. Pressure at the compressor outlet (which for practical purposes will be the same as that in the receiver) is called the working pressure and is used to specify the compressor. Pressure at the operating point is called, not surprisingly, the operating pressure and is used to specify valves, actuators and other operating devices.

Care should be taken in specifying the volume of gas a compressor is required to deliver. Expression 3.1 shows the volume of a given mass of gas to be highly dependent on pressure and temperature. Delivery volume of a compressor is defined in terms of gas at normal atmospheric conditions. Two standards known as *standard temperature and pressures* (STP) are commonly used, although differences between them are small for industrial users.

The *technical normal condition* is:

$$P = 0.98 \text{ bar absolute}, T = 20°C$$

and the *physical normal condition* is:

$$P = 1.01 \text{ bar absolute}, T = 0°C$$

The term *normal temperature and pressure* (NTP) is also used.

Required delivery volume of a compressor (in M^3 min^{-1} or ft^3 min^{-1}, according to the units used) may be calculated for the actuators at the various operating positions (with healthy safety margins to allow for leakage) but care must be taken to ensure this total volume is converted to STP condition before specifying the required compressor delivery volume.

A compressor delivery volume can be specified in terms of its theoretical volume (swept volume multiplied by rotational speed) or effective volume which includes losses. The ratio of these two volumes is the efficiency. Obviously the effective volume should be used in choosing a compressor (with, again, a safety margin for leakage). Required power of the motor driving the compressor is dependent on working pressure and delivery volume, and may be determined from expressions 2.2 and 2.5. Allowance must be made for the cyclic on/off operation of the compressor with the motor being sized for on load operation and not averaged over a period of time.

Piston compressors

Piston compressors are by far the most common type of compressor, and a basic single cylinder form is shown in Figure 3.3. As the piston descends during the inlet stroke (Figure 3.3a), the inlet valve opens and air is drawn into the cylinder. As the piston passes the bottom of the stroke, the inlet valve closes and the exhaust valve opens allowing air to be expelled as the piston rises (Figure 3.3b)

Figure 3.3 implies that the valves are similar to valves in an internal combustion engine. In practice, spring-loaded valves are used, which open and close under the action of air pressure across them. One common type uses a 'feather' of spring steel which moves above the inlet or output port, as shown in Figure 3.3c.

A single cylinder compressor gives significant pressure pulses at the outlet port. This can be overcome to some extent by the use of a large receiver, but more often a multicylinder compressor is used. These are usually classified as vertical or

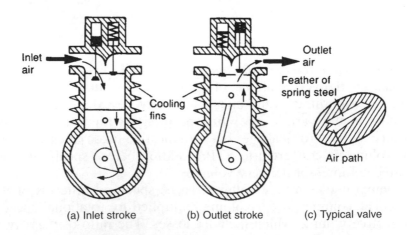

(a) Inlet stroke   (b) Outlet stroke   (c) Typical valve

Figure 3.3 *Single cylinder compressor*

horizontal in-line arrangements and the more compact V, Y or W constructions.

A compressor which produces one pulse of air per piston stoke (of which the example of Figure 3.3 is typical) is called a single-acting compressor. A more even air supply can be obtained by the double acting action of the compressor in Figure 3.4, which uses two sets of valves and a crosshead to keep the piston rod square at all times. Double-acting compressors can be found in all configurations described earlier.

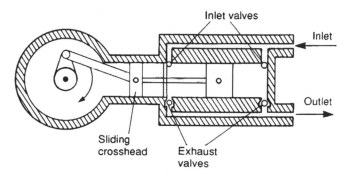

Figure 3.4 *Double-acting compressor*

Piston compressors described so far go direct from atmospheric to required pressure in a single operation. This is known as a single stage compressor. The general gas law (expression 1.19) showed compression of a gas to be accompanied by a significant rise in gas temperature. If the exit pressure is above about 5 bar in a single-acting compressor, the compressed air temperature can rise to over 200°C and the motor power needed to drive the compressor rises accordingly.

For pressures over a few bar it is far more economical to use a multistage compressor with cooling between stages. Figure 3.5 shows an example. As cooling (undertaken by a device called an intercooler) reduces the volume of the gas to be compressed at the second stage there is a large energy saving. Normally two stages are used for pneumatic pressures of 10 to 15 bar, but multistage compressors are available for pressures up to around 50 bar.

Multistage compressors can be manufactured with multicylinders as shown in Figure 3.5 or, more compactly, with a single cylinder and a double diameter piston as shown in Figure 3.6.

There is contact between pistons and air, in standard piston compressors, which may introduce small amounts of lubrication oil

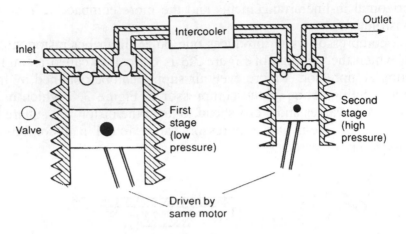

Figure 3.5   *Two-stage compressor*

from the piston walls into the air. This very small contamination may be undesirable in food and chemical industries. Figure 3.7 shows a common way of giving a totally clean supply by incorporating a flexible diaphragm between piston and air.

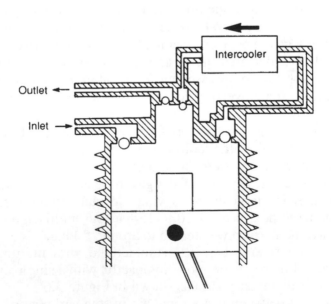

Figure 3.6   *Combined two-stage compressor*

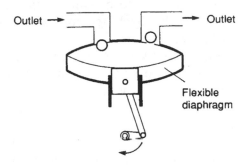

Figure 3.7 *Diaphragm compressor, used where air must not be contaminated*

Screw compressors

Piston compressors are used where high pressures (> 20 bar) and relatively low volumes (< 10,000 m³ hr⁻¹) are needed, but are mechanically relatively complex with many moving parts. Many applications require only medium pressure (< 10 bar) and medium flows (around 10,000 m³ hr⁻¹). For these applications, rotary compressors have the advantage of simplicity, with fewer moving parts rotating at a constant speed, and a steady delivery of air without pressure pulses.

One rotary compressor, known as the dry rotary screw compressor, is shown in Figure 3.8 and consists of two intermeshing rotating screws with minimal (around 0.05 mm) clearance. As the

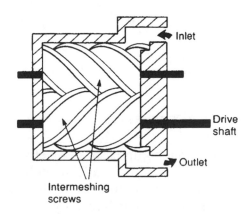

Figure 3.8 *Dry screw rotary compressor*

screws rotate, air is drawn into the housing, trapped between the screws and carried along to the discharge port, where it is delivered in a constant pulse-free stream.

Screws in this compressor can be synchronised by external timing gears. Alternatively one screw can be driven, the second screw rotated by contact with the drive screw. This approach requires oil lubrication to be sprayed into the inlet air to reduce friction between screws, and is consequently known as a wet rotary screw compressor. Wet screw construction though, obviously introduces oil contamination into the air which has to be removed by later oil separation units.

Rotary compressors

The vane compressor, shown in Figure 3.9 operates on similar principles to the hydraulic vane pump described in Chapter 2, although air compressors tend to be physically larger than hydraulic pumps. An unbalanced design is shown, balanced versions can also be constructed. Vanes can be forced out by springs or, more commonly, by centrifugal force.

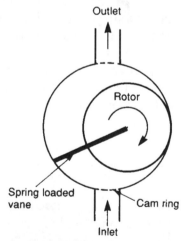

Figure 3.9   *Vane compressor*

A single stage vane compressor can deliver air at up to 3 bar, a much lower pressure than that available with a screw or piston compressor. A two-stage vane compressor with large low pressure and smaller high pressure sections linked by an intercooler allows pressures up to 10 bar to be obtained.

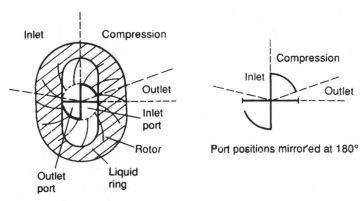

Figure 3.10 *Liquid ring compressor*

Figure 3.10 shows a variation on the vane compressor called a liquid ring compressor. The device uses many vanes rotating inside an eccentric housing and contains a liquid (usually water) which is flung out by centrifugal force to form a liquid ring which follows the contour of the housing to give a seal with no leakage and minimal friction. Rotational speed must be high (typically 3000 rpm) to create the ring. Delivery pressures are relatively low at around 5 bar.

The lobe compressor of Figure 3.11 (often called a Roots blower) is often used when a positive displacement compressor is needed with high delivery volume but low pressure (typically 1–2 bar). Operating pressure is mainly limited by leakage between rotors and housing. To operate efficiently, clearances must be very small, and wear leads to a rapid fall in efficiency.

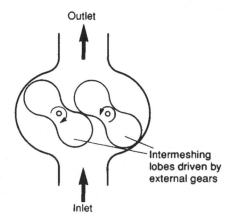

Figure 3.11 *Lobe compressor*

Dynamic compressors

A large volume of air (up to 5000 m^3 min^{-1}) is often required for applications such as pneumatic conveying (where powder is carried in an air stream), ventilation or where air itself is one component of a process (e.g. combustion air for gas/oil burners). Pressure in these applications is low (at most a few bar) and there is no need for a positive displacement compressor.

Large volume low pressure air is generally provided by dynamic compressors known as blowers. They can be subdivided into centrifugal or axial types, shown in Figure 3.12. Centrifugal blowers (Figure 3.12a) draw air in then fling it out by centrifugal force. A high shaft rotational speed is needed and the volume to input power ratio is lower than any other type of compressor.

An axial compressor comprises a set of rotating fan blades as shown in Figure 3.12b. These produce very large volumes of air, but at low pressure (less than one bar). They are primarily used for ventilation, combustion and process air.

Output pressures of both types of dynamic compressor can be lifted by multistage compressors with intercoolers between stages. Diffuser sections reduce air entry velocity to subsequent stages, thereby converting air kinetic energy to pressure energy.

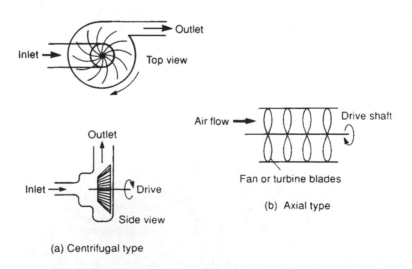

Figure 3.12   *Non-positive displacement compressors (Blowers)*

Positive displacement compressors use oil to lubricate the close machined parts and to maintain the air seal. Dynamic compressors have no such need, and consequently deliver very clean air.

Air receivers and compressor control

An air receiver is used to store high pressure air from the compressor. Its volume reduces pressure fluctuations arising from changes in load and from compressor switching.

Air coming from the compressor will be warm (if not actually hot!) and the large surface area of the receiver dissipates this heat to the surrounding atmosphere. Any moisture left in the air from the compressor will condense out in the receiver, so outgoing air should be taken from the receiver top.

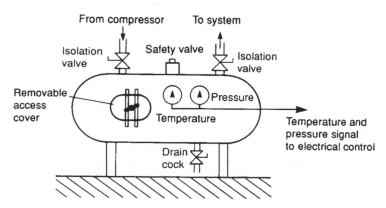

Figure 3.13   *Compressed air receiver*

Figure 3.13 shows essential features of a receiver. They are usually of cylindrical construction for strength, and have a safety relief valve to guard against high pressures arising from failure of the pressure control scheme. Pressure indication and, usually, temperature indication are provided, with pressure switches for control of pressure and high temperature switches for remote alarms.

A drain cock allows removal of condensed water, and access via a manhole allows cleaning. Obviously, removal of a manhole cover is hazardous with a pressurised receiver, and safety routines must be defined and followed to prevent accidents.

Control of the compressor is necessary to maintain pressure in the receiver. The simplest method of achieving this is to start the compressor when receiver pressure falls to some minimum pressure, and stop the compressor when pressure rises to a satisfactory level again, as illustrated in Figure 3.14. In theory two pressure switches are required (with the motor start pressure lower than the motor stop pressure) but, in practice, internal hysteresis in a typical switch allows one pressure switch to be used. The pressure in the receiver cycles between the start and stop pressure settings.

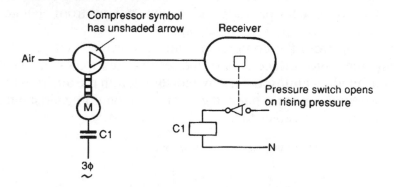

Figure 3.14   *Receiver pressure control via motor start/stop*

In Figure 3.15 another method of pressure control is shown, where the compressor runs continuously and an exhaust valve is fitted to the compressor outlet. This valve opens when the required pressure is reached. A non-return valve prevents air returning from the receiver. This technique is known as exhaust regulation.

Compressors can also be controlled on the inlet side. In the example of Figure 3.16, an inlet valve is held open to allow the compressor to operate, and is closed when the air receiver has

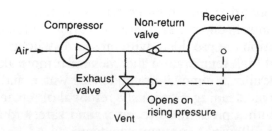

Figure 3.15   *Receiver pressure control using compressor outlet valve*

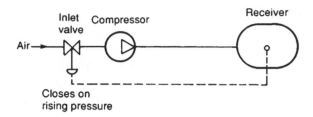

Figure 3.16 *Receiver pressure control using compressor inlet valve*

reached the desired pressure, (the compressor then forms a near vacuum on its inlet side).

The valves in Figure 3.15 and 3.16 can be electrically-operated solenoid valves controlled by pressure switches, or can be pneumatic valves controlled directly by receiver pressure.

The control method is largely determined by flow rates from receiver to the load(s) and the capacity of the compressor. If the compressor has significant spare capacity, for example, start/stop control is commonly used.

If compressor capacity and load requirements are closely matched, on the other hand, short start/stop cycling may cause premature wear in the electrical starter for the compressor motor. In this situation, exhaust or inlet regulation is preferred.

Air receiver size is determined by load requirements, compressor capacity, and allowable pressure deviations in the receiver. With the compressor stopped, Boyle's law (expression 1.17) gives the pressure decay for a given volume of air delivered from a given receiver at a known pressure. For example, if a receiver of 10 cubic metres volume and a working pressure of 8 bar delivers 25 cubic metres of air (at STP) to a load, pressure in the receiver falls to approximately 5.5 bar.

With the compressor started, air pressure rises at a rate again given by expression 1.17 (with the air mass in the receiver being *increased* by the difference between the air delivered by the compressor and that removed by the load).

These two calculations give the cycle time of the compressor when combined with settings of the cut-in and drop-out pressure switches. If this is unacceptably rapid, say less than a few minutes, then a larger receiver is required. Manufacturers of pneumatic equipment provide nomographs which simplify these calculations.

An air receiver is a pressure vessel and as such requires regular visual and volumetric pressure tests. Records should be kept of the tests.

Air treatment

Atmospheric air contains moisture in the form of water vapour. We perceive the amount of moisture in a given volume of air as the humidity and refer to days with a high amount of water vapour as 'humid' or 'sticky', and days with low amounts of water vapour as 'good drying days'. The amount of water vapour which can be held in a given volume depends on temperature but does *not* depend on pressure of air in that volume. One cubic metre at 20°C, for example, can hold 17 grams of water vapour. The amount of water vapour which can be held in a given volume of air rises with temperature as shown in Figure 3.17.

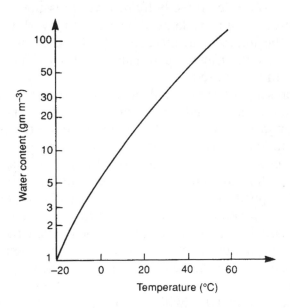

Figure 3. 17   *Moisture content curve*

If a given volume of air contains the maximum quantity of water vapour possible at the air temperature, the air is said to be *saturated* (and we would perceive it as sticky because sweat could not evaporate from the surface of the skin). From Figure 3.17, air containing 50 grams of water vapour per cubic metre at 40°C is saturated.

Moisture content of unsaturated air is referred to by relative humidity, which is defined as:

$$\text{Relative humidity} = \frac{\text{water content per cubic metre}}{\text{maximum water content per cubic metre}} \times 100\%. \quad (3.2)$$

Air containing 5 grams of water vapour per cubic metre of air at 20°C has, from Figure 3.17, a relative humidity of 30%.

Relative humidity is dependent on both temperature and pressure of the air. Suppose air at 30°C contains 20 grams of water vapour. From Figure 3.17 this corresponds to 67% humidity. If the air is allowed to cool to 20°C it can only hold 17 grams of water vapour and is now saturated (100% relative humidity). The excess 3 grams condenses out as liquid water. If the air is cooled further to 10°C, a further 8 grams condenses out.

The temperature at which air becomes saturated is referred to as the 'dew point'. Air with 17.3 grams of water vapour per cubic metre has, for example, a dew point of 20°C.

To see the effect of pressure on relative humidity, we must remember the amount of water vapour which can be held in a given volume is fixed (assuming a constant temperature). Suppose a cubic metre of air at atmospheric pressure (0 bar gauge or 1 bar absolute) at 20°C contains 6 grams of water vapour (corresponding to 34% relative humidity). If we wish to increase air pressure while maintaining its temperature at 20°C, we must compress it. When the pressure is 1 bar gauge (or 2 bar absolute) its volume is 0.5 cubic metres, which can hold 8.6 grams of water vapour, giving us 68% relative humidity. At 2 bar gauge (3 bar absolute) the volume is 0.33 cubic metres, which can hold 5.77 grams of water vapour. With 6 grams of water vapour in our air, we have reached saturation and condensation has started to occur.

It follows that relative humidity rises quickly with increasing pressure, and even low atmospheric relative humidity leads to saturated air and condensation at the pressures used in pneumatic systems (8–10 bar). Water droplets resulting from this condensation can cause many problems. Rust will form on unprotected steel surfaces, and the water may mix with oil (necessary for lubrication) to form a sticky white emulsion, which causes valves to jam and blocks the small piping used in pneumatic instrumentation systems. In extreme cases water traps can form in pipe loops.

When a compressed gas expands suddenly there is a fall of temperature (predicted by expression 1.19). If the compressed air has a high water content, a rapid expansion at exhaust ports can be accompanied by the formation of ice as the water condenses out and freezes.

Stages of air treatment

Air in a pneumatic system must be clean and dry to reduce wear and extend maintenance periods. Atmospheric air contains many harmful impurities (smoke, dust, water vapour) and needs treatment before it can be used.

In general, this treatment falls into three distinct stages, shown in Figure 3.18. First, inlet filtering removes particles which can damage the air compressor. Next, there is the need to dry the air to reduce humidity and lower the dew point. This is normally performed between the compressor and the receiver and is termed primary air treatment.

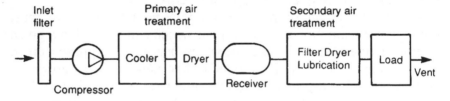

Figure 3.18   *Three stages of air treatment*

The final treatment is performed local to the duties to be performed, and consists of further steps to remove moisture and dirt and the introduction of a fine oil mist to aid lubrication. Not surprisingly this is generally termed secondary air treatment.

Filters

Inlet filters are used to remove dirt and smoke particles before they can cause damage to the air compressor, and are classified as dry filters with replaceable cartridges (similar to those found in motor car air filters) or wet filters where the incoming air is bubbled through an oil bath then passed through a wire mesh filter. Dirt particles became attached to oil droplets during the bubbling process and are consequently removed by the wire mesh.

Both types of filter require regular servicing: replacement of the cartridge element for the dry type; cleaning for the wet type. If a filter is to be cleaned, it is essential the correct detergent is used. Use of petrol or similar petrochemicals can turn an air compressor into an effective diesel engine – with severe consequences.

Filters are classified according to size of particles they will stop. Particle size is measured in SI units of micrometres (the older metric term *microns is* still common) one micrometre (1 μm) being 10^{-6} metre or 0.001 millimetre. Dust particles are generally larger than 10 μm, whereas smoke and oil particles are around 1 μm. A filter can have a nominal rating (where it will block 98% of particles of the specified size) or an absolute rating (where it blocks 100% of particles of the specified size).

Microfilters with removable cartridges passing air from the centre to the outside of the cartridge case will remove 99.9% of particles down to 0.01 μm, the limit of normal filtration. Coarse filters, constructed out of wire mesh and called strainers, are often used as inlet filters. These are usually specified in terms of the mesh size which approximates to particle size in micrometres as follows:

Mesh size	μm
325	30
550	10
750	6

Air dryers

An earlier section described how air humidity and dew point are raised by compression. Before air can be used, this excess moisture has to be removed to bring air humidity and dew point to reasonable levels.

In bulk air systems all that may be required is a simple after-cooler similar to the intercoolers described earlier, followed by a separator unit where the condensed water collects and can be drained off.

Figure 3.19a shows a typical water trap and separator. Air flow through the unit undergoes a sudden reversal of direction and a deflector cone swirls the air (Figure 3.19b). Both of these cause heavier water particles to be flung out to the walls of the separator and to collect in the trap bottom from where they can be drained. Water traps are usually represented on circuit diagrams by the symbol of Figure 3.19c.

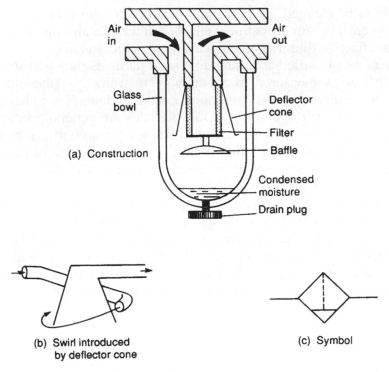

(a) Construction

(b) Swirl introduced by deflector cone

(c) Symbol

Figure 3.19   Air filter and water trap

Dew point can be lowered further with a refrigerated dryer, the layout of which is illustrated in Figure 3.20. This chills the air to just above 0°C condensing almost all the water out and collecting the condensate in the separator. Efficiency of the unit is improved with a second heat exchanger in which cold dry air leaving the dryer pre-chills incoming air. Air leaving the dryer has a dew point similar to the temperature in the main heat exchanger.

Refrigerated dryers give air with a dew point sufficiently low for most processes. Where absolutely dry air is needed, chemical dryers must be employed. Moisture can be removed chemically from air by two processes.

In a deliquescent dryer, the layout of which is shown in Figure 3.21, a chemical agent called a desiccant is used. This absorbs water vapour and slowly dissolves to form a liquid which collects at the bottom of the unit where it can be drained. The dessicant material is used up during this process and needs to be replaced at regular intervals. Often deliquescent dryers are referred to as absorbtion dryers. a term that should not be confused with the next type of dryer.

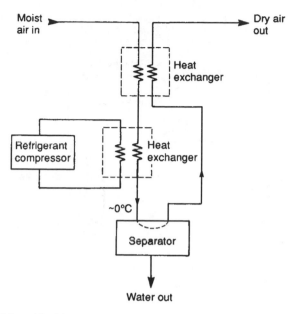

Figure 3.20 *Refrigerated dryer*

An adsorption dryer collects moisture on the sharp edges of a granular material such as silicon dioxide, or with materials which can exist in hydrated *and* de-hydrated states (the best known is copper sulphate but more efficient compounds are generally used). Figure 3.22 shows construction of a typical adsorption dryer. Moisture in the adsorption material can be released by heating, so

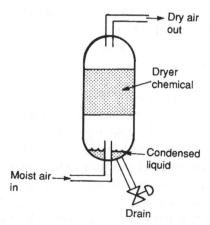

Figure 3.21 *Deliquescent dryer*

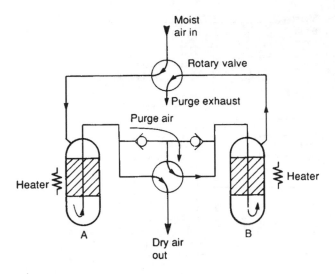

Figure 3.22   *Adsorption dryer*

two columns are used. At any time, one column is drying the air while the other is being regenerated by heating and the passage of a low purge air stream. As shown, column A dries the air and column B is being regenerated. The rotary valves are operated automatically at regular intervals by a time clock. For obvious reasons adsorption dryers are often referred to as regenerative dryers.

Lubricators

A carefully controlled amount of oil is often added to air immediately prior to use to lubricate moving parts (process control pneumatics are the exception as they usually require dry unlubricated air). This oil is introduced as a fine mist, but can only be added to thoroughly clean and dry air or a troublesome sticky emulsion forms. It is also difficult to keep the oil mist-laden air in a predictable state in an air receiver, so oil addition is generally performed as part of the secondary air treatment.

The construction of a typical lubricator is shown with its symbol in Figure 3.23. The operation is similar to the principle of the petrol air mixing in a motor car carburettor As air enters the lubricator its velocity is increased by a venturi ring causing a local reduction in pressure in the upper chamber. The pressure differential between lower and upper chambers causes oil to be drawn up a

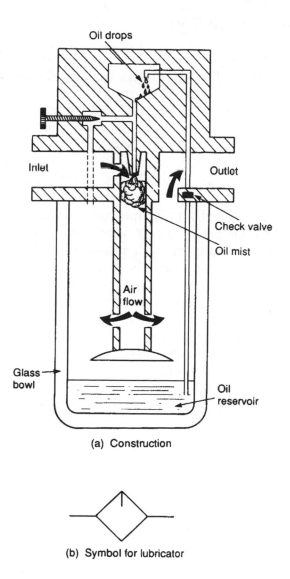

(a) Construction

(b) Symbol for lubricator

Figure 3.23 *Lubricator*

riser tube, emerging as a spray to mix with the air. The needle valve adjusts the pressure differential across the oil jet and hence the oil flow rate.

The air–oil mixture is forced to swirl as it leaves the central cylinder causing excessively large oil particles to be flung out of the air stream.

Pressure regulation

Flow velocities in pneumatic systems can be quite high, which can lead to significant flow-dependent pressure drops between the air receiver and the load.

Generally, therefore, air pressure in the receiver is set higher than the required load pressure and pressure regulation is performed local to loads to keep pressure constant regardless of flow. Control of air pressure in the receiver was described in an earlier section. This section describes various ways in which pressure is locally controlled.

There are essentially three methods of local pressure control, illustrated in Figure 3.24. Load A vents continuously to atmosphere. Air pressure is controlled by a pressure regulator which simply restricts air flow to the load. This type of regulator requires some minimum flow to operate. If used with a dead-end load which draws no air, the air pressure will rise to the main manifold pressure. Such regulators, in which air must pass through the load, are called non-relieving regulators.

Load B is a dead-end load, and uses a pressure regulator which vents air to atmosphere to reduce pressure. This type of regulator is called a three-port (for the three connections) or relieving regulator. Finally, load C is a large capacity load whose air volume requirements are beyond the capacity of a simple in-line regulator. Here a pressure control loop has been constructed com-

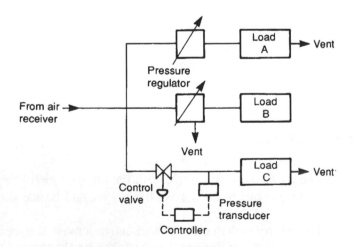

Figure 3.24   *Three types of pressure regulator*

prising pressure transducer, electronic controller and separate vent valve. This technique can also be used if the pressure regulating valve cannot be mounted locally to the point at which the pressure is to be controlled.

Relief valves

The simplest pressure regulating device is the relief valve shown in Figure 3.25. This is not, in fact, normally used to control pressure but is employed as a backup device should the main pressure control device fail. They are commonly fitted, for example, to air receivers.

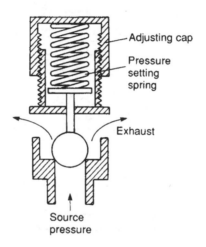

Figure 3.25   *Relief valve*

A ball valve is held closed by spring tension, adjustable to set the relief pressure. When the force due to air pressure exceeds the spring tension, the valve cracks open releasing air and reducing the pressure. Once cracked, flow rate is a function of excess pressure; an increase in pressure leading to an increase in flow. A relief valve is specified by operating pressure range, span of pressure between cracking and full flow, and full flow rate. Care is needed in specifying a relief valve because in a fault condition the valve may need to pass the entire compressor output.

A relief valve has a flow/pressure relationship and self-seals itself once the pressure falls below the cracking pressure. A pure safety valve operates differently. Once a safety valve cracks, it opens fully to discharge all the pressure in the line or receiver, and it does not automatically reclose, needing manual resetting before the system can be used again.

Non-relieving pressure regulators

Figure 3.26 shows construction of a typical non-relieving pressure regulator. Outlet pressure is sensed by a diaphragm which is pre-loaded by a pressure setting spring. If outlet pressure is too low, the spring forces the diaphragm and poppet down, opening the valve to admit more air and raise outlet pressure.

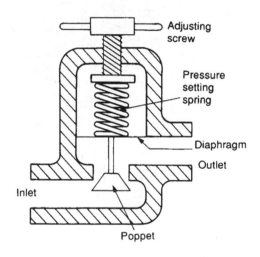

Figure 3.26   *Non-relieving pressure regulator*

   If the outlet pressure is too high, air pressure forces the diaphragm up, reducing air flow and causing a reduction in air pressure as air vents away through the load. In a steady state the valve will balance with the force on the diaphragm from the outlet pressure just balancing the preset force on the spring.

Relieving pressure regulators

A relieving regulator is shown in Figure 3.27. Outlet pressure is sensed by a diaphragm preloaded with an adjustable pressure

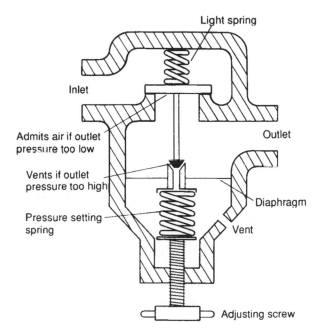

Figure 3.27 *Relieving pressure regulator*

setting spring. The diaphragm rises if the outlet pressure is too high, and falls if the pressure is too low.

If outlet pressure falls, the inlet poppet valve is pushed open admitting more air to raise pressure. If the outlet pressure rises, the diaphragm moves down closing the inlet valve and opening the central vent valve to allow excess air to escape from the load thereby reducing pressure.

In a steady state the valve will balance; *dithering* between admitting and venting small amounts of air to keep load pressure at the set value.

Both the regulators in Figure 3.26 and 3.27 are simple pressure regulators and have responses similar to that shown in Figure 3.28, with outlet pressure decreasing slightly with flow. This droop in pressure can be overcome by using a pilot-operated regulator, shown in Figure 3.29.

Outlet pressure is sensed by the pilot diaphragm, which compares outlet pressure with the value set by the pressure setting spring. If outlet pressure is low the diaphragm descends, while if outlet pressure is high the diaphragm rises.

Inlet air is bled through a restriction and applied to the top of the

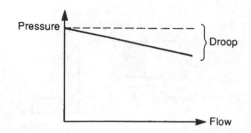

Figure 3.28 *Response of simple pressure regulators*

main diaphragm. This space can, however, be vented to the exit side of the valve by the small ball valve.

If outlet pressure is low, the pilot diaphragm closes the ball valve causing the main diaphragm to be pushed down and more air to be admitted to the load.

If outlet pressure is high, the pilot diaphragm opens the ball valve and the space above the main diaphragm de-pressurises. This

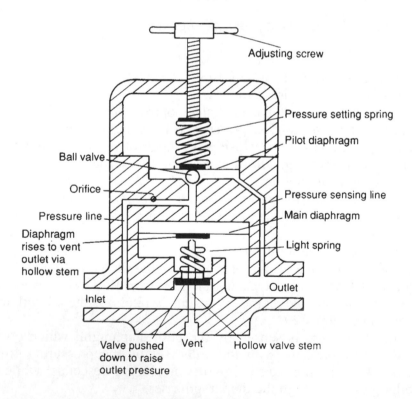

Figure 3.29 *Pilot-operated regulator*

causes the main diaphragm to rise, opening the central vent allowing air to escape from the load and pressure to be reduced.

Action of pilot diaphragm and inlet air bleed approximates to integral action, giving a form of P + I (proportional plus integral) control. In the steady state, the outlet pressure equals the set pressure and there is no pressure droop with increasing flow.

Service units

In pneumatic systems a moisture separator, a pressure regulator, a pressure indicator, a lubricator and a filter are all frequently required, local to a load or system. This need is so common that combined devices called *service units* are available. Individual components comprising a service unit are shown in Figure 3.30a, while the composite symbol of a service unit is shown in Figure 3.30b.

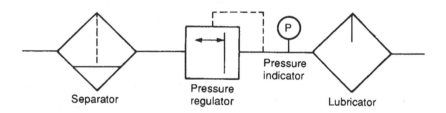

(a) Symbols for individual components

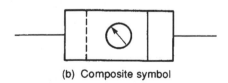

(b) Composite symbol

Figure 3.30   *The service unit*

4

Control valves

Pneumatic and hydraulic systems require control valves to direct and regulate the flow of fluid from compressor or pump to the various load devices. Although there are significant practical differences between pneumatic and hydraulic devices (mainly arising from differences in operating pressures and types of seals needed for gas or liquid) the operating principles and descriptions are very similar.

Although valves are used for many purposes, there are essentially only two types of valve. An infinite position valve can take up any position between open and closed and, consequently, can be used to modulate flow or pressure. Relief valves described in earlier chapters are simple infinite position valves.

Most control valves, however, are only used to allow or block flow of fluid. Such valves are called finite position valves. An analogy between the two types of valve is the comparison between an electric light dimmer and a simple on/off switch. Connections to a valve are termed 'ports'. A simple on/off valve therefore has two ports. Most control valves, however, have four ports shown in hydraulic and pneumatic forms in Figure 4.1.

In both the load is connected to ports labelled A, B and the pressure supply (from pump or compressor) to port P. In the hydraulic valve, fluid is returned to the tank from port T. In the pneumatic valve return air is vented from port R.

Figure 4.2 shows internal operation of valves. To extend the ram, ports P and B are connected to deliver fluid and ports A and T connected to return fluid. To retract the ram, ports P and A are connected to deliver fluid and ports B and T to return fluid.

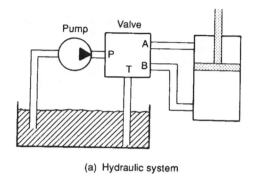

(a) Hydraulic system

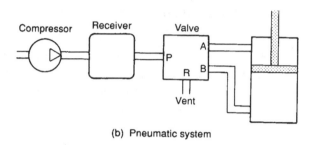

(b) Pneumatic system

Figure 4.1 *Valves in a pneumatic and hydraulic system*

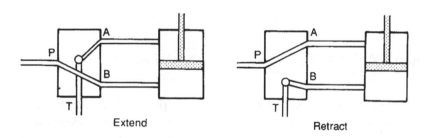

Extend   Retract

Figure 4.2 *Internal valve operation*

Another consideration is the number of control positions. Figure 4.3 shows two possible control schemes. In Figure 4.3a, the ram is controlled by a lever with two positions; extend or retract. This valve has two control positions (and the ram simply drives to one end or other of its stroke). The valve in Figure 4.3b has three positions; extend, off, retract. Not surprisingly the valve in Figure 4.3a is called a two position valve, while that in Figure 4.3b is a three position valve.

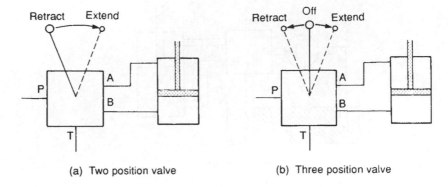

(a) Two position valve   (b) Three position valve

Figure 4.3   *Valve control positions*

Finite position valves are commonly described as a port/position valve where *port* is the number of ports and *position* is the number of positions. Figure 4.3a therefore illustrates a 4/2 valve, and Figure 4.3b shows a 4/3 valve. A simple block/allow valve is a 2/2 valve.

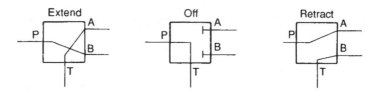

Figure 4.4   *Possible valve action for a 4/3 valve*

The numbers of ports and positions does not, however, completely describe the valve. We must also describe its action. Figure 4.4 shows one possible action for the 4/3 valve of Figure 4.3b. Extend and retract connections are similar, but in the off position ports P and T are connected–unloading the pump back to the tank without need of a separate loading valve, while leaving the ram locked in position. (This approach could, of course, only be used where the pump supplies one load). Other possible arrangements may block all four ports in the off position (to maintain pressure), or connect ports A, B and T (to leave the ram free in the off position). A complete valve description thus needs number of ports, number of positions *and* the control action.

Graphic symbols

Simple valve symbols have been used so far to describe control actions. From the discussions in the previous section it can be seen that control actions can easily become too complex for representation by sketches showing how a valve is constructed.

A set of graphic symbols has therefore evolved (similar, in principle, to symbols used on electrical circuit diagrams). These show component function without showing the physical construction of each device. A 3/2 spool valve and a 3/2 rotary valve with the same function have the same symbol; despite their totally different constructions.

Symbols are described in various national documents; DIN24300, BS2917, ISO1219 and the new ISO5599, CETOP RP3 plus the original American JIC and ANSI symbols. Differences between these are minor.

A valve is represented by a square for each of its switching positions. Figure 4.5a thus shows the symbol of a two position valve, and Figure 4.5b a three position valve. Valve positions can be represented by letters a, b, c and so on, with 0 being used for a central neutral position.

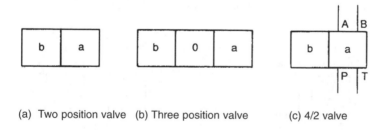

(a) Two position valve   (b) Three position valve   (c) 4/2 valve

Figure 4.5 *Basis of graphic symbols*

Ports of a valve are shown on the outside of boxes in normal unoperated or initial position. Four ports have been added to the two position valve symbol shown in Figure 4.5c. Designations given to ports are normally:

Port	Designation
Working lines	A, B, C and so on
Pressure (power) supply	P
Exhaust/Return	R, S, T and so on (T normally used for hydraulic systems, R and S for pneumatic systems)
Control (Pilot) Lines	Z, Y, X and so on

ISO 5599 proposes to replace these letters with numbers, a retrograde step in the author's opinion.

Arrow-headed lines represent direction of flow. In Figure 4.6a, for example fluid is delivered from port P to port A and returned from port B to port T when the valve is in its normal state a. In state b, flow is reversed. This valve symbol corresponds to the valve represented in Figures 4.2 and 4.3a.

Shut off positions are represented by **T**, as shown by the central position of the valve in Figure 4.6b, and internal flow paths can be represented as shown in Figure 4.6c. This latter valve, incidentally, vents the load in the off position.

In pneumatic systems, lines commonly vent to atmosphere directly at the valve, as shown by port R in Figure 4.6d.

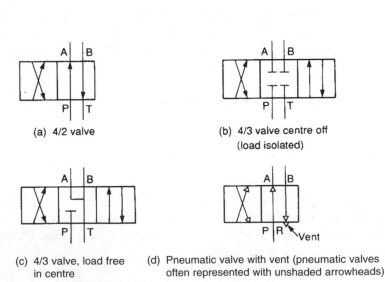

(a) 4/2 valve

(b) 4/3 valve centre off
(load isolated)

(c) 4/3 valve, load free
in centre

(d) Pneumatic valve with vent (pneumatic valves
often represented with unshaded arrowheads)

Figure 4.6   *Valve symbols*

Figure 4.7a shows symbols for the various ways in which valves can be operated. Figure 4.7b thus represents a 4/2 valve operated by a pushbutton. With the pushbutton depressed the ram extends. With the pushbutton released, the spring pushes the valve to position a and the ram retracts.

Actuation symbols can be combined. Figure 4.7c represents a solenoid-operated 4/3 valve, with spring return to centre.

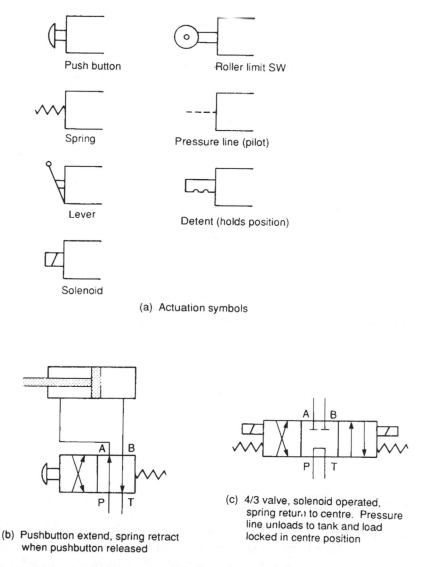

Push button

Roller limit SW

Spring

Pressure line (pilot)

Lever

Detent (holds position)

Solenoid

(a) Actuation symbols

(b) Pushbutton extend, spring retract when pushbutton released

(c) 4/3 valve, solenoid operated, spring return to centre. Pressure line unloads to tank and load locked in centre position

Figure 4.7 *Complete valve symbols*

Infinite position valve symbols are shown in Figure 4.8. A basic valve is represented by a single square as shown in Figure 4.8a, with the valve being shown in a normal, or non-operated, position. Control is shown by normal actuation symbols: in Figure 4.8b, for example, the spring pushes the valve right decreasing flow, and pilot pressure pushes the valve left increasing flow. This represents a pressure relief valve which would be connected into a hydraulic system as shown in Figure 4.8c.

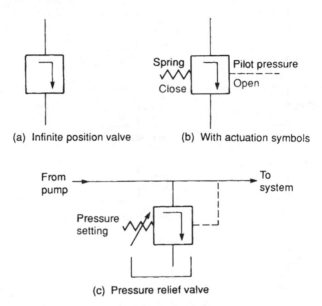

(a) Infinite position valve

(b) With actuation symbols

(c) Pressure relief valve

Figure 4.8   *Infinite position valve symbols*

Types of control valve

There are essentially three types of control valve; poppet valves spool valves and rotary valves.

Poppet valves

In a poppet valve, simple discs, cones or balls are used in conjunction with simple valve seats to control flow. Figure 4.9 shows the construction and symbol of a simple 2/2 normally-closed valve, where depression of the pushbutton lifts the ball off its seat and

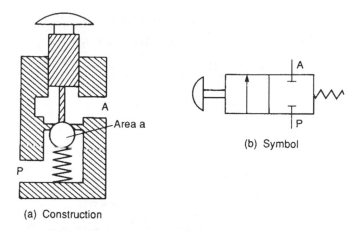

(a) Construction

(b) Symbol

Figure 4.9 *Simple 2/2 poppet valve*

allows fluid to flow from port P to port A. When the button is released, spring and fluid pressure force the ball up again closing the valve.

Figure 4.10 shows the construction and symbol of a disc seal 3/2 poppet. With the pushbutton released, ports A and R are linked via the hollow pushbutton stem. If the pushbutton is pressed, port R is first sealed, then the valve disc pushed down to open the valve and connect ports P and A. As before, spring and fluid pressure from port P closes the valve.

The valve construction and symbol shown in Figure 4.11 is a poppet changeover 4/2 valve using two stems and disc valves. With the pushbutton released, ports A and R are linked via the hollow left-hand stem and ports P and B linked via the normally-open right hand disc valve. When the pushbutton is pressed, the link between ports A and R is first closed, then the link between P and B closed. The link between A and P is next opened, and finally the link between B and R opened. When the pushbutton is released, air and spring pressure puts the valve back to its original state.

Poppet valves are simple, cheap and robust, but it is generally simpler to manufacture valves more complicated than those shown in Figure 4.11 by using spool valves. Further, a major disadvantage of poppet valves is the force needed to operate them. In the poppet valve of Figure 4.10, for example, the force required on the pushbutton to operate the valve is P × a newtons. Large capacity valves need large valve areas, leading to large operating force. The high pressure in hydraulic systems thus tends to prevent use of simple

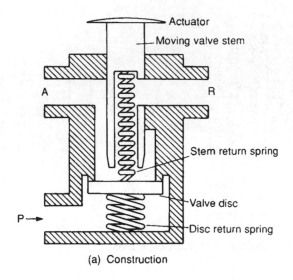

(a) Construction

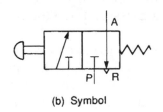

(b) Symbol

Figure 4.10   *A 3/2 poppet valve*

poppet valves and they are, therefore, mainly found in low pressure pneumatic systems.

Spool valves

Spool (or slide) valves are constructed with a spool moving horizontally within the valve body, as shown for the 4/2 valve in Figure 4.12. Raised areas called 'lands' block or open ports to give the required operation.

The operation of a spool valve is generally balanced. In the valve construction in Figure 4.12b, for example, pressure is applied to opposing faces D and E and low tank pressure to faces F and G. There is no net force on the spool from system pressure, allowing the spool to be easily moved.

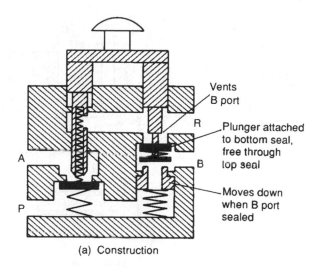

(a) Construction

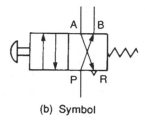

(b) Symbol

Figure 4.11 *A 4/2 poppet valve*

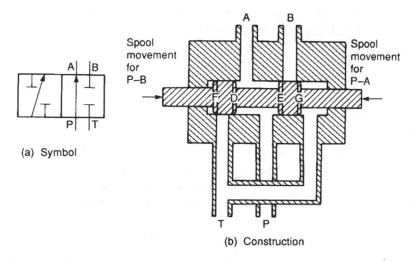

Figure 4.12 *Two-way spool valve*

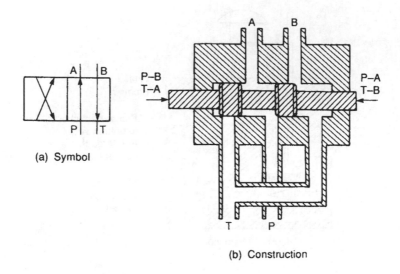

(a) Symbol

(b) Construction

Figure 4.13   *Four-way spool valve*

Figure 4.13 is a changeover 4/2 spool valve. Comparison of the valves shown in Figures 4.12 and 4.13 shows they have the same body construction, the only difference being the size and position of lands on the spool. This is a major cost-saving advantage of spool valves; different operations can be achieved with a common body and different spools. This obviously reduces manufacturing costs.

Figure 4.14 shows various forms of three position changeover valves; note, again, these use one body with different functions achieved by different land patterns.

Spool valves are operated by shifting the spool. This can be achieved by button, lever or striker, or remotely with a solenoid. Self-centring can easily be provided if springs are mounted at the end of the spool shaft.

Solenoid-operated valves commonly work at 24 V DC or 110 V AC. Each has its own advantages and disadvantages. A DC power supply has to be provided for 24 V DC solenoids, which, in large systems, is substantial and costly. Operating current of a 24 V solenoid is higher than a 110 V solenoid's. Care must be taken with plant cabling to avoid voltage drops on return legs if a common single line return is used.

Current through a DC solenoid is set by the winding resistance. Current in an AC solenoid, on the other hand, is set by the inductance of the windings, and this is usually designed to give a high inrush current followed by low holding current. This is achieved by using the core of the solenoid (linked to the spool) to raise the coil inductance when the spool has moved. One side effect of this is that a jammed spool results in a permanent high current which can damage the coil or the device driving it. Each and every AC solenoid should be protected by an individual fuse. DC solenoids do not suffer from this characteristic. A burned out DC solenoid coil is almost unknown.

Whatever form of solenoid is used it is very useful when fault finding to have local electrical indication built into the solenoid plug top. This allows a fault to be quickly identified as either an electrical or hydraulic problem. Fault finding is discussed further in Chapter 8.

A solenoid can exert a pull or push of about 5 to 10 kg. This is adequate for most pneumatic spool valves, but is too low for direct operation of large capacity hydraulic valves. Here pilot operation must be used, a topic discussed later.

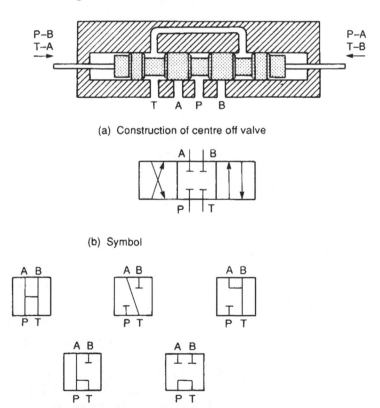

(a) Construction of centre off valve

(b) Symbol

(c) Common centre position connections

Figure 4.14   *Three position four-way valves*

Rotary valves

Rotary valves consist of a rotating spool which aligns with holes in the valve casing to give the required operation. Figure 4.15 shows the construction and symbol of a typical valve with centre off action.

Rotary valves are compact, simple and have low operating forces. They are, however, low pressure devices and are consequently mainly used for hand operation in pneumatic systems.

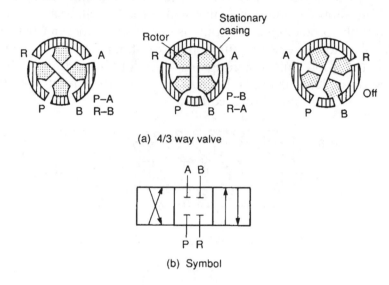

(a) 4/3 way valve

(b) Symbol

Figure 4.15   *Rotary valve*

Pilot-operated valves

With large capacity pneumatic valves (particularly poppet valves) and most hydraulic valves, the operating force required to move the valve can be large. If the required force is too large for a solenoid or manual operation, a two-stage process called pilot operation is used.

The principle is shown in Figure 4.16. Valve 1 is the main operating valve used to move a ram. The operating force required to move the valve, however, is too large for direct operation by a solenoid, so a second smaller valve 2, known as the pilot valve, has been added to allow the main valve to be operated

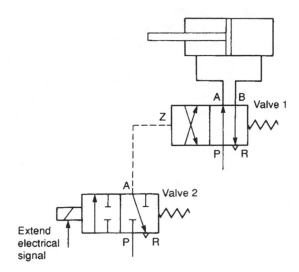

Figure 4.16 *Pilot-operated valve*

by system pressure. Pilot pressure lines are normally shown dotted in circuit diagrams, and pilot ports on main valves are denoted Z, Y, X and so on.

In Figure 4 16, pilot port Z is depressurised with the solenoid de-energised, and the ram is retracted. When the solenoid is energised valve 2 changes over, pressurising Z; causing valve 1 to energise and the ram to extend.

Although pilot operation can be achieved with separate valves it is more usual to use a pilot/main valve assembly manufactured as a complete ready made unit. Figure 4.17 shows the operation of a pilot-operated 3/2 pneumatic valve. The solenoid operates the small pilot valve directly. Because this valve has a small area, a low operating force is required. The pilot valve applies line pressure to the top of the control valve causing it to move down, closing the exhaust port. When it contacts the main valve disc there are two forces acting on the valve stem. The pilot valve applies a downwards force of $P \times D$, where P is the line pressure and D is the area of the control valve. Line pressure also applies an upwards force $P \times E$ to the stem, where E is the area of the main valve. The area of the control valve, D, is greater than area of the main valve E, so the downwards force is the larger and the valve opens.

When the solenoid de-energises, the space above the control

valve is vented. Line and spring pressure on the main valve causes the valve stem to rise again, venting port A.

A hydraulic 4/2 pilot-operated spool valve is shown in Figure 4.18. The ends of the pilot spool in most hydraulic pilot-operated valves are visible from outside the valve. This is useful from a maintenance viewpoint as it allows the operation of a valve to be checked. In extreme cases the valve can be checked by pushing the pilot spool directly with a suitably sized rod (welding rod is ideal!). Care must be taken to check solenoid states on dual solenoid valves before attempting manual operation. Overriding an energised AC solenoid creates a large current which may damage the coil, (or blow the fuse if the solenoid has correctly installed protection).

Check valves

Check valves only allow flow in one direction and, as such, are similar in operation to electronic diodes. The simplest construction is the ball and seat arrangement of the valve in Figure 4.19a, commonly used in pneumatic systems. The right angle construction in Figure 4.19b is better suited to the higher pressures of a hydraulic

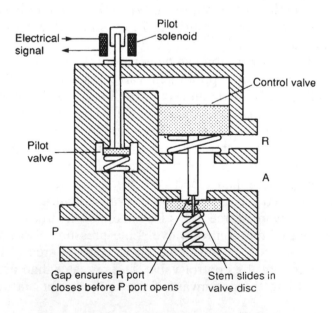

Figure 4.17   *Construction of a pilot-operated 3/2 valve*

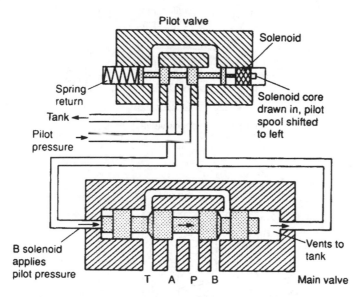

Pilot valve

Solenoid

Spring
return

Tank

Pilot
pressure

Solenoid core
drawn in, pilot
spool shifted
to left

B solenoid
applies
pilot pressure

Vents to
tank

T   A   P   B

Main valve

(a) Construction: power applied to solenoid has moved pilot
spool to left. This applies pilot pressure to left hand end of main
spool, shifting spool to right and connecting P & B ports

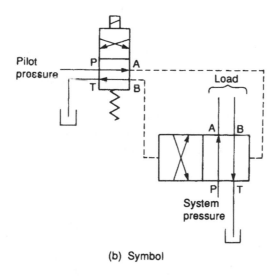

(b) Symbol

Figure 4.18 *Pilot-operated valve*

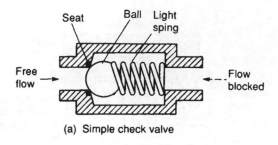

(a) Simple check valve

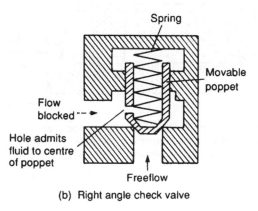

(b) Right angle check valve

Figure 4.19   *Check valves*

system. Free flow direction is normally marked with an arrow on the valve casing.

A check valve is represented by the graphic symbols in Figure 4.20. The symbol in Figure 4.20a is rather complex and the simpler symbol in Figure 4.20b is more commonly used.

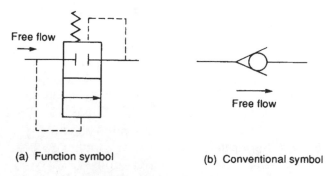

(a) Function symbol

(b) Conventional symbol

Figure 4.20   *Check valve symbols*

Figure 4.21 illustrates several common applications of check valves. Figure 4.21a shows a combination pump, used where an application requires large volume and low pressure, or low volume and high pressure. A typical case is a clamp required to engage quickly (high volume and low pressure) then grip (minimal volume but high pressure). Pump 1 is the high volume and low pressure pump, and pump 2 the high pressure pump. In high volume mode both pumps deliver to the system, pump 1 delivering through the check valve V_3. When high pressure is required, line pressure at X rises operating unloading valve V_1 via pilot port Z taking pump 1 off load. Pump 2 delivers the required pressure set by relief valve V_2, with the check valve preventing fluid leaking back to pump 1 and V_1.

Figure 4.21b shows a hydraulic circuit with a pressure storage device called an accumulator (described in a later chapter). Here a check valve allows the pump to unload via the pressure regulating valve, while still maintaining system pressure from the accumulator.

A spring-operated check valve requires a small pressure to open (called the cracking pressure) and acts to some extent like a low pressure relief valve. This characteristic can be used to advantage. In Figure 4.21c pilot pressure is derived before a check valve, and in Figure 4.21d a check valve is used to protect a blocked filter by diverting flow around the filter when pressure rises. A check valve is also included in the tank return to prevent fluid being sucked out of the tank when the pump is turned off.

Pilot-operated check valves

The cylinder in the system in Figure 4.22 should, theoretically, hold position when the control valve is in its centre, off, position. In practice, the cylinder will tend to creep because of leakage in the control valve.

Check valves have excellent sealage in the closed position, but a simple check valve cannot be used in the system in Figure 4.22 because flow is required in both directions. A pilot-operated check is similar to a basic check valve but can be held open permanently by application of an external pilot pressure signal.

There are two basic forms of pilot-operated check valves, shown in Figure 4.23. They operate in a similar manner to basic check valves, but with pilot pressure directly opening the valves. In the 4C valve shown in Figure 4.23a, inlet pressure assists the pilot. The

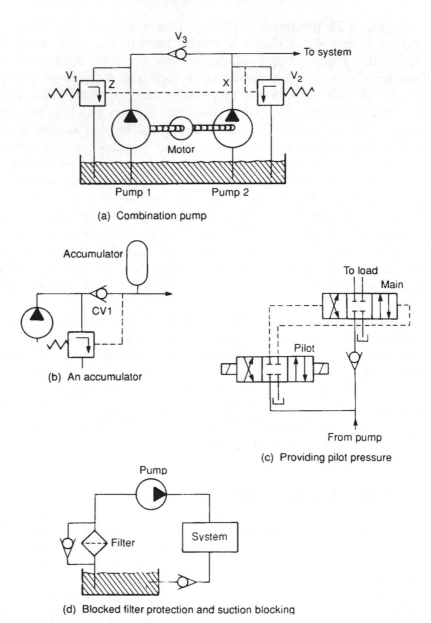

(a) Combination pump

(b) An accumulator

(c) Providing pilot pressure

(d) Blocked filter protection and suction blocking

Figure 4.21   *Check valve applications*

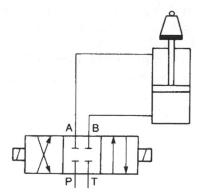

Figure 4.22 *System requiring a check valve. In the off position the load 'creeps'*

symbol of a pilot-operated check valve is shown in Figure 4.23c.

The cylinder application of Figure 4.22 is redrawn with pilot-operated check valves in Figure 4.23d. The pilot lines are connected to the pressure line feeding the other side of the cylinder. For any cylinder movement, one check valve is held open by flow (operating as a normal check valve) and the other is held open by pilot pressure. For no required movement, both check valves are closed and the cylinder is locked in position.

Restriction check valves

The speed of a hydraulic or pneumatic actuator can be controlled by adjusting the rate at which a fluid is admitted to, or allowed out from, a device. This topic is discussed in more detail in Chapter 5 but a speed control is often required to be direction-sensitive and this requires the inclusion of a check valve.

A restriction check valve (often called a throttle relief valve in pneumatics) allows full flow in one direction and a reduced flow in the other direction. Figure 4.24a shows a simple hydraulic valve and Figure 4.24b a pneumatic valve. In both, a needle valve sets restricted flow to the required valve. The symbol of a restriction check valve is shown in Figure 4.24c.

Figure 4.24d shows a typical application in which the cylinder extends at full speed until a limit switch makes, then extend further at low speed. Retraction is at full speed.

A restriction check valve V_2 is fitted in one leg of the cylinder. With the cylinder retracted, limit-operated valve V_3 is open allowing free flow of fluid from the cylinder as it extends. When the striker plate on the cylinder ram hits the limit, valve V_3 closes and flow out of the cylinder is now restricted by the needle valve setting of valve V_2. In the reverse direction, the check valve on valve V_2 opens giving full speed of retraction.

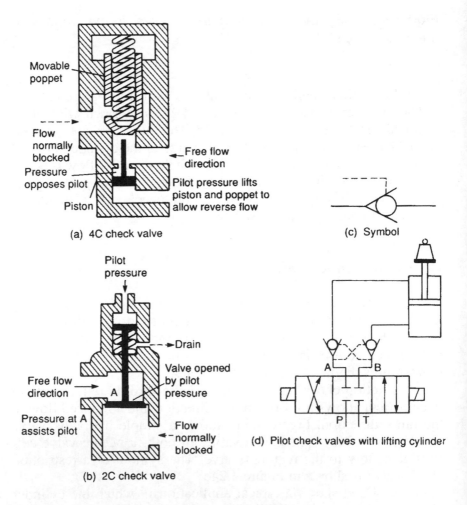

(a) 4C check valve

(c) Symbol

(b) 2C check valve

(d) Pilot check valves with lifting cylinder

Figure 4.23   *Pilot-operated check valves*

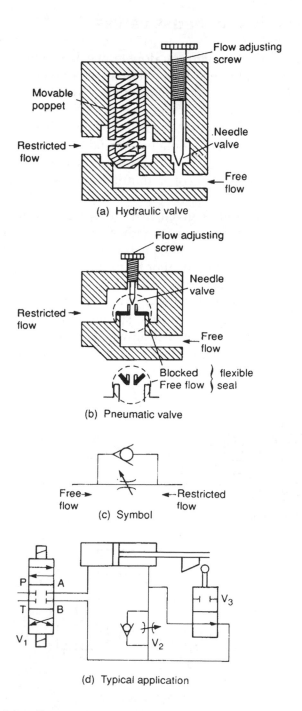

(a) Hydraulic valve

(b) Pneumatic valve

(c) Symbol

(d) Typical application

Figure 4.24 *Restriction check valve*

Shuttle and fast exhaust valves

A shuttle valve, also known as a double check valve, allows pressure in a line to be obtained from alternative sources. It is primarily a pneumatic device and is rarely found in hydraulic circuits.

Construction is very simple and consists of a ball inside a cylinder, as shown in Figure 4.25a. If pressure is applied to port X, the ball is blown to the right blocking port Y and linking ports X and A. Similarly, pressure to port Y alone connects ports Y and A and blocks port X. The symbol of a shuttle valve is given in Figure 4.25b.

A typical application is given in Figure 4.25c, where a spring return cylinder is operated from either of two manual stations. Isolation between the two stations is provided by the shuttle valve. Note a simple T-connection cannot be used as each valve has its A port vented to the exhaust port.

A fast exhaust valve (Figure 4.26) is used to vent cylinders quickly. It is primarily used with spring return (single-acting) pneumatic cylinders. The device shown in Figure 4.26a consists of a movable disc which allows port A to be connected to

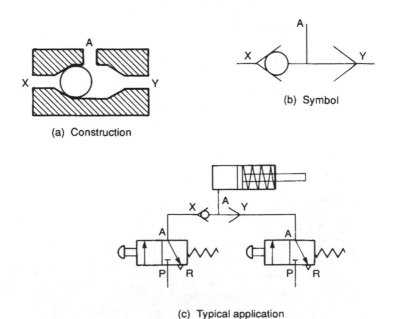

(a) Construction

(b) Symbol

(c) Typical application

Figure 4.25   *Pneumatic shuttle valve*

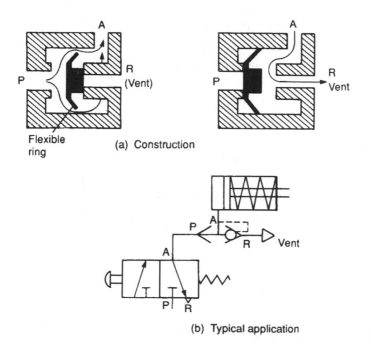

(a) Construction

(b) Typical application

Figure 4.26 *Fast exhaust valve*

pressure port P or large exhaust port R. It acts like, and has the same symbol as, a shuttle valve. A typical application is shown in Figure 4.26b.

Fast exhaust valves are usually mounted local to, or directly onto, cylinders and speed up response by avoiding any delay from return pipes and control valves. They also permit simpler control valves to be used.

Sequence valves

The sequence valve is a close relative of the pressure relief valve and is used where a set of operations are to be controlled in a pressure related sequence. Figure 4.27 shows a typical example where a workpiece is pushed into position by cylinder 1 and clamped by cylinder 2.

Sequence valve V_2 is connected to the extend line of cylinder 1. When this cylinder is moving the workpiece, the line pressure is low, but rises once the workpiece hits the end stop. The sequence valve opens once its inlet pressure rises above a preset level.

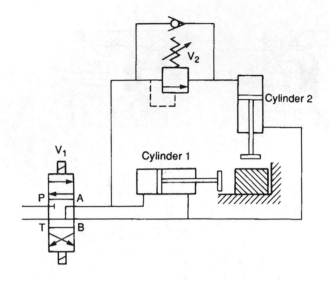

Figure 4.27   *Sequence valve*

Cylinder 2 then operates to clamp the workpiece. A check valve across V_2 allows both cylinders to retract together.

Time delay valves

Pneumatic time delay valves (Figure 4.28) are used to delay operations where time-based sequences are required. Figure 4.28a shows construction of a typical valve. This is similar in construction to a 3/2 way pilot-operated valve, but the space above the main valve is comparatively large and pilot air is only allowed in via a flow-reducing needle valve. There is thus a time delay between application of pilot pressure to port Z and the valve operation, as shown by the timing diagram in Figure 4.28b. The time delay is adjusted by the needle valve setting.

The built-in check valve causes the reservoir space above the valve to vent quickly when pressure at Z is removed to give no delay off.

The valve shown in Figure 4.28 is a normally-closed delay-on valve. Many other time delay valves (delay-off, delay on/off, normally-open) can be obtained. All use the basic principle of the air reservoir and needle valve.

The symbol of a normally-dosed time delay valve is shown in Figure 4.28c.

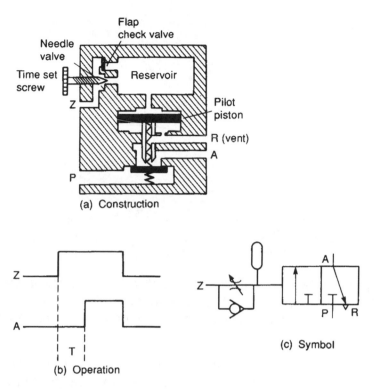

Figure 4.28 *Pneumatic time delay valve*

Proportional Valves

The solenoid valves described so far act, to some extent, like an electrical switch, i.e. they can be On or Off. In many applications it is required to remotely control speed, pressure or force via an electrical signal. This function is provided by proportional valves.

A typical two position solenoid is only required to move the spool between 0 and 100% stroke against the restoring force of a spring. To ensure predictable movement between the end positions the solenoid must also increase its force as the spool moves to ensure the solenoid force is larger than the increasing opposing spring force at all positions.

A proportional valve has a different design requirement. The spool position can be set anywhere between 0% and 100% stroke by varying the solenoid current. To give a predictable response the solenoid must produce a force which is dependent solely on the

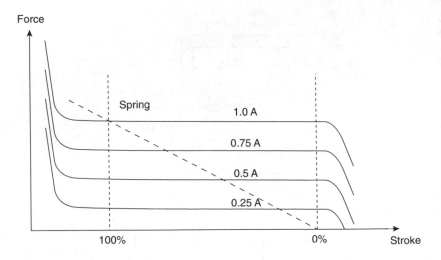

Figure 4.29   *The relationship between coil current force and stroke for a proportional valve solenoid. Note the flat part of the curve and the linear relationship between current and force*

current and not on the spool position, i.e. the force for a given current must be constant over the full stroke range. Furthermore, the force must be proportional to the current.

Figure 4.29 shows a typical response. The force from the solenoid is opposed by the force from a restoring spring, and the spool will move to a position where the two forces are equal. With a current of 0.75 A, for example, the spool will move to 75% of its stroke.

The spool movement in a proportional valve is small; a few mm stroke is typical. The valves are therefore very vulnerable to stiction, and this is reduced by using a 'wet' design which immerses the solenoid and its core in hydraulic fluid.

A proportional valve should produce a fluid flow which is proportional to the spool displacement. The spools therefore use four triangular metering notches in the spool lands as shown on Figure 4.30. As the spool is moved to the right, port A will progressively link to the tank and port B to the pressure line.

The symbol for this valve is also shown. Proportional valves are drawn with parallel lines on the connection sides of the valve block on circuit diagrams.

Figure 4.30 gives equal flow rates to both A and B ports. Cylinders have different areas on the full bore and annulus sides

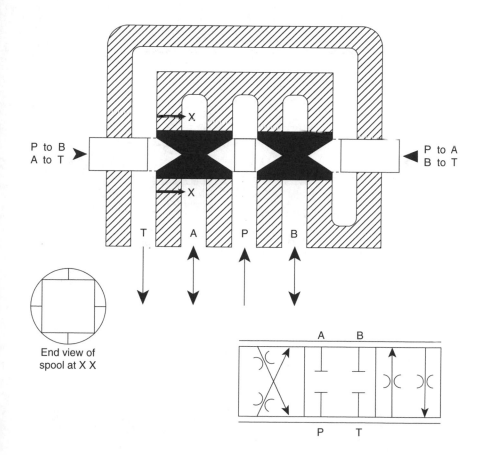

Figure 4.30 *Construction and symbol for a proportional valve. When used with a cylinder with 2:1 full bore to annulus area ratio, half the V cutouts will be provided on one of the P lands*

(see Figure 5.4). To achieve equal speeds in both directions, the notches on the lands must have different areas. With a 2:1 cylinder ratio, half the number of notches are used on one side.

Figure 4.31 shows the construction and symbol for a restricted centre position valve. Here the extended notches provide a restricted (typically 3%) flow to tank from the A and B ports when the valve is in the centre position.

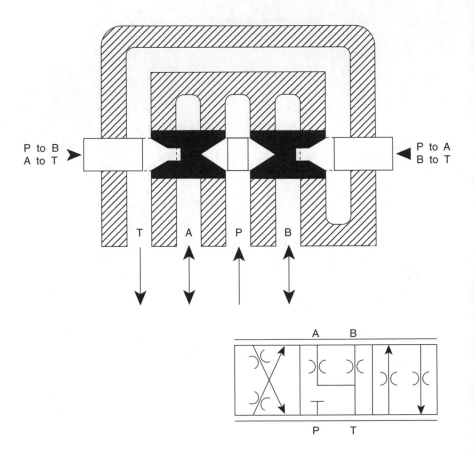

Figure 4.31 *Construction and symbol for a proportional valve with A and B ports linked to tank in the null position*

So far we have assumed the spool position is determined by the balance between the force from the solenoid and the restoring force from a spring. Whilst this will work for simple applications, factors such as hydraulic pressure on the spool and spring ageing mean the repeatability is poor. Direct solenoid/spring balance is also not feasible with a pilot/main spool valve. What is really required is some method of position control of the spool.

To achieve this, the spool position must be measured. Most valves use a device called a Linear Variable Differential

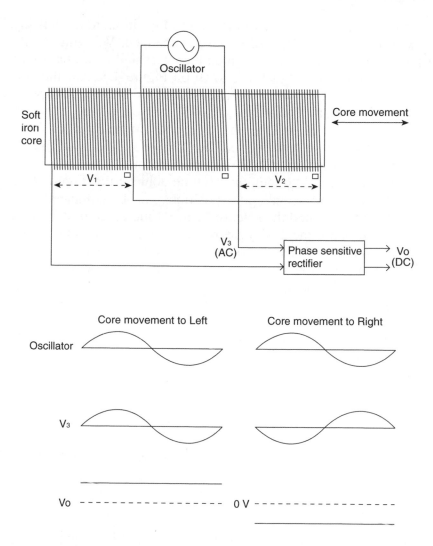

Figure 4.32 *The use of an LVDT to give position control of the valve spool. The LVDT can be connected to the pilot or main spool. (a) The circuit arrangement of the LVDT and phase sensitive rectifier. (b) Output signals for core displacement to left and right*

Transformer (or LVDT) shown on Figure 4.32a. The LVDT consists of a soft iron core whose position is to be measured surrounded by three electrical windings. A high frequency (typically a few kHz) AC signal is applied to the centre winding which induces voltages into the other two windings. When the core is central, V_1 and V_2 are equal but opposite in phase giving zero volts at V_3.

If the core moves away from the central position, to the left say, V_2 will decrease, but V_1 will remain unchanged. V_3 (which is the difference between V_1 and V_2) thus increases and is in phase with the driving oscillator signal as shown on Figure 4.32b. If the core moves to the right V_3 will also increase, but will now be anti-phase to the driving signal. The amplitude of V_3 is proportional to the distance the core moves, and the phase depends on the direction. V_3 is connected to a phase sensitive rectifier to give a bi-polar DC output signal V_o proportional to the core displacement.

A position control system can now be achieved as Figure 4.33. The demanded and actual spool positions can be compared by a position controller, and the solenoid current increased or decreased automatically until the position error is zero. In a pilot/main valve the position feedback will be taken from the main spool

The spool position is determined by the solenoid current. A typical solenoid will operate over a range of about 0 to 1 amp. Power dissipation in the current controller is $V \times I$ watts where V is the volts drop and I the current. Maximum dissipation occurs at half current (0.5 A) which, with a typical 24 V supply, gives 12 watts. This implies substantial, bulky (and hence expensive), power transistors.

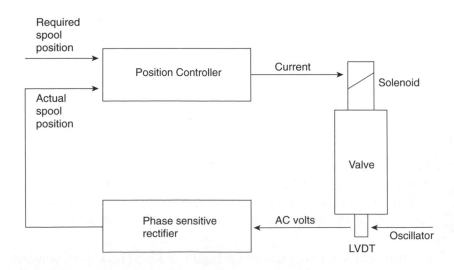

Figure 4.33   *Position of the spool in a proportional valve with an LVDT and a phase sensitive rectifier. In many systems the oscillator, LVDT and phase sensitive rectifier are now included in the valve itself*

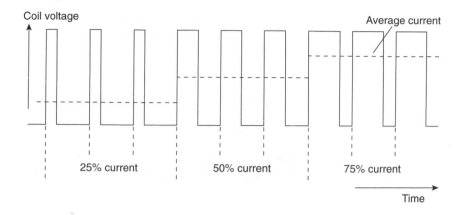

Figure 4.34 *Pulse Width Modulation (PWM) used to control solenoid current with minimal power dissipation in the output transistors*

Current control is usually performed with Pulse Width Modulation (PWM) shown in Figure 4.34. Here the current is turned rapidly On and Off with the On/Off ratio determining the mean current. The control circuit is either turned fully on (low voltage drop, high current but low dissipation) or fully off (high voltage drop, zero current, again low dissipation). Because the dissipation is low, smaller and cheaper transistors can be used.

Proportional valves operate with small forces from the solenoid and rely on small deflections of the spool. They are hence rather vulnerable to stiction which causes the valve to ignore small changes in demanded spool position. The effect is made worse if the valve spool is held in a fixed period for a period of time, allowing the spool to settle. Dirt in the oil also encourages stiction as small dirt particles will increase the probability of the spool sticking.

A high frequency (typically a few kHz), signal is therefore added to the command signal as Figure 4.35. This is too fast for the valve to follow, but the small movement prevents the spool from staying in a fixed position. This action, called Dither, is normally factory set on the electronic control card described below.

It is not possible for a proportional valve to totally shut off flow in the centre, null, position unless the spool is manufactured with a small deadband as Figure 4.36. The result is a non-linear response

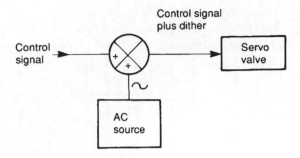

Figure 4.35   *Using dither to reduce stiction. The dither frequency and amplitude are normally a factory preset on the electronic control card*

between demanded spool position and the resultant flow. In many cases this is of no concern, but if full reversing control is required the deadband may be a problem.

Most electronic control cards thus include a deadband compensation. This adds an adjustable offset to the reference signal in each direction effectively allowing the width of the deadband region to be controlled.

Sudden changes of speed imply large accelerations which in turn imply large forces since $F = ma$ where F is the force, m is the mass

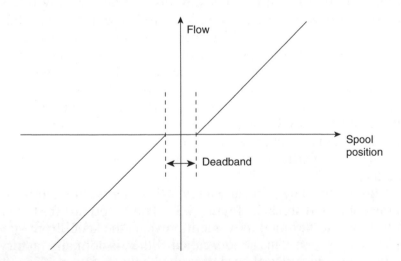

Figure 4.36   *To prevent flow in the null (centre) position most proportional valves have a small deadband as shown. This can be offset by a Deadband adjustment on the controlling card*

and *a* the acceleration. At best, sudden speed changes will result in noise from the system. More probably, however, the step forces will result in eventual damage and failure of piping, pumps and actuators. Most proportional valve control circuits therefore include methods by which the acceleration and deceleration can be controlled as shown on Figure 4.37a and b. Here four ramp rates, two for acceleration and two for deceleration, soften the impact of the stepped demanded input signal. These ramp rates can be pre-set, usually by trim potentiometers on the electronic control card described below.

Figure 4.37 allows independent adjustment of acceleration and deceleration in all four quadrants (A,B,C and D). In simpler, (and hence cheaper), arrangements there may be two adjustable ramp rates for acceleration and deceleration (i.e. A and D are equal and B and C are equal), or two ramp rates according to slope sign (i.e. A and C are equal and B and D are equal). In the simplest case there is only a single adjustable ramp rate (i.e. A, B, C and D are all equal).

A proportional valve must be used with some form of electronic control. Usually this is provided by a single card per valve. Cards can be mounted onto a back plate or, more usually, in a 19 inch rack. Figure 4.38 shows a typical card schematic.

Electronic cards for proportional valves usually run on a single 24 volt power supply, and require a current of around 1 to 2 amps; not insignificant when several cards are being used on the same project. The tolerance on the supply volts is usually quite wide, typically 20 to 30 V is quoted. Diode D1 on the card protects against inadvertent supply reversal.

An on board power supply produces the multiple supply rails needed by the card circuit; +15 V, +10 V, −10 V and −15 V are common, with 5 V on microprocessor based cards. The +10 V and −10 V supplies are brought out to card terminal as supplies for a manually adjusted control potentiometer.

The Enable input allows current to pass to the valve solenoids. To enable the card, this must be connected to +24 V. This input can be used for safety critical functions such as emergency stops, overtravel limits, safety gates etc.

The valve reference can come in many forms; the card illustrated uses three. First is a voltage signal with a range from +10 V (solenoid A fully open) to −10 V (solenoid B fully open). This signal range is normally used with a manual control potentiometer. The second signal accepts the standard instrumentation signal of 4–20 mA to cover the same valve range. Current signals are less

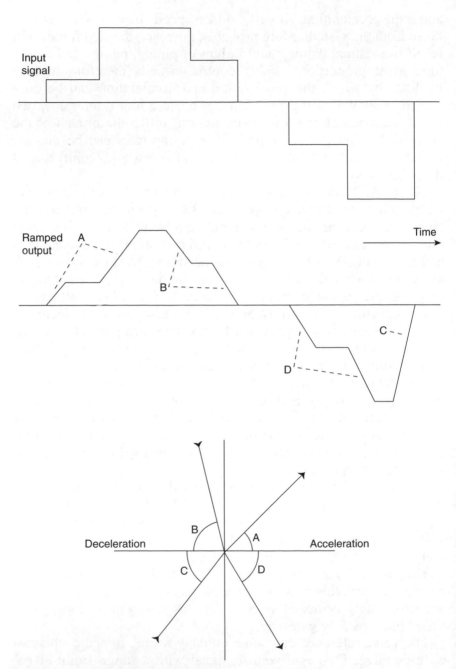

Figure 4.37 *Ramped response. Four quadrant operation is shown, single ramp rate and two quadrant operation are more common. (a) Effect of applying ramps with four quadrant feature. (b) Definition of the four quadrants*

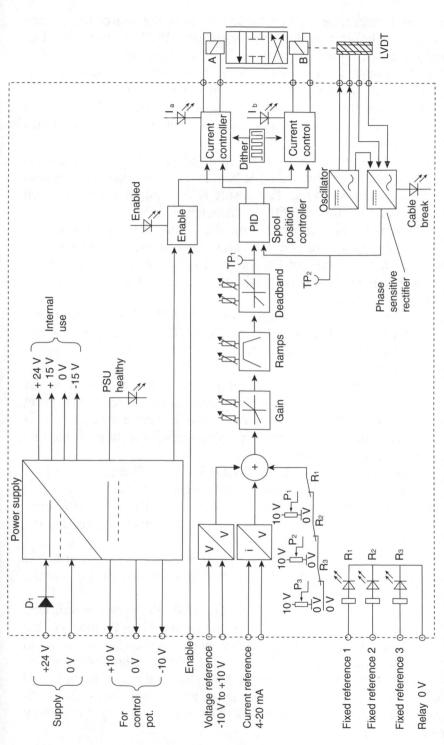

Figure 4.38 *Block Diagram of a typical electronic card for a proportional valve. In many systems the oscillator, LVDT and phase sensitive rectifier are all included in the valve itself*

prone to interference on long cable runs from the source to the card if a valve is being remotely controlled by a PLC or computer. The final reference comes from three fixed settings on potentiometers P1 to P3 mounted on the card itself. These are selected by digital signals which energize relays R1 to R3. The resulting reference is the *sum* of all three. In practice only one will be used, the others being zero. On some cards the source is selected by small switches on the card.

The resulting reference is then adjusted for gain, ramp rate (two quadrants shown on this example, single and four quadrants are also common) and deadband. The result is a required spool position which can be monitored with a voltmeter on TP1.

This setpoint is compared with the actual spool position, also available for monitoring on TP2, and the error used by a three term (proportional integral and derivative, PID) controller to adjust the current to solenoids A and B. Dither is added to the current signals to reduce stiction.

The spool position is monitored by an LVDT, fed from an oscillator on the card. The signal from the LVDT is turned into a DC signal by a phase sensitive rectifier and fed back to the PID spool position controller.

Extensive monitoring and diagnostic facilities are built into the card. The desired and actual spool positions are a crucial test point as these show if the valve is responding to the reference signal. This provides a natural break point for diagnostics, as it shows if the reference is being received.

Another useful test points are LEDs I_a and I_b. These glow with an intensity which is proportional to the solenoid current. If the valve sticks, for example, one LED will shine brightly as the PID controller sends full current to try to move the valve and reduce the error between TP1 and TP2.

Other LED's show the correct operation of the power supply, the state of the Enable signal, the selected fixed speed (if used) and a cable break fault from the LVDT.

Figure 4.38 is based on conventional electronic amplifiers. Increasingly microprocessors are being used, and although the operation is identical in function, it is performed by software. Serial communications, (RS232, RS485 or Fieldbus standards such as Profibus), is becoming common for adjusting the reference and reading the valve status. The settings of gains, ramp rates, fixed references etc. can be set remotely and easily changes by a computer or PLC control system.

With microprocessor based cards, stepper motors are often used to position the spool via a screw thread. This removes the need to balance a solenoid force against a spring force and combines the spool positioning actuator and feedback in the same device.

As electronics becomes smaller there is also a tendency to move the PID controller, current controllers and LVDT circuit into the valve head itself, i.e. everything to the right of TP1. Here the card simply provides a spool reference and a 24 V supply to the valve.

The valves described so far are directional valves, allowing flow to be controlled to and from a load. A proportional valve can also be used to control pressure. The principle is shown on Figure 4.39.

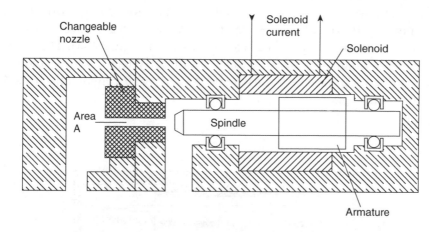

Figure 4.39 *Proportional pressure control valve. The pressure is given by the force produced by the solenoid divided by the area*

The solenoid spindle is aligned with a nozzle connected to the pressure line. For oil to pass from the pressure line back to tank, the force resulting from the fluid pressure must exceed the solenoid force. The relief valve will thus pass fluid back to the tank if the pressure force exceeds the solenoid force, and the pressure will be maintained at

$$\text{Pressure} = \frac{\text{solenoid force}}{\text{nozzle area}}$$

The solenoid force is directly proportional to the solenoid current, so the pressure is also directly proportional to the current. The range of the relief valve is set by the nozzle area, and manufacturers supply nozzle inserts with different areas.

The circuit of Figure 4.39 can only handle a small fluid flow, so a practical valve will incorporate a proportional valve pilot stage linked to a main stage in a similar manner to the manually set spring operated relief valve of Figure 2.6b.

Servo valves

Servo valves are a close relative of the proportional valve and are based on an electrical torque motor which produces a small deflection proportional to the electrical current through its coil. They commonly use feedback between the main and pilot spools to give precise control. A typical device is shown on Figure 4.40. This consists of a small pilot spool connected directly to the torque motor. The pilot spool moves within a sliding sleeve, mechanically linked to the main spool.

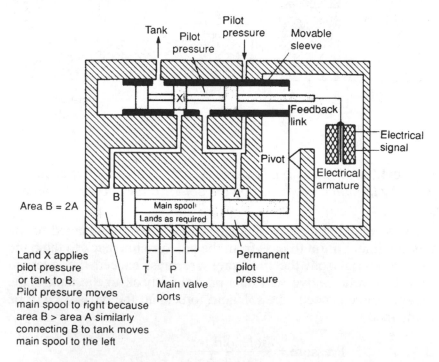

Figure 4.40   *Two-stage servo valve*

The right-hand end of the main spool is permanently connected to the pilot pressure line, but because of the linkage rod its area is reduced to an annulus of area A. Pressure at the left-hand end of the spool is controlled by the pilot valve. There is no area restriction at this end, and the valve is designed such that the spool has an area of 2A.

If the same pressure P is applied to both ends, the spool experiences a left force of $P \times A$ and a right force of $2P \times A$ causing a net force of $P \times A$ to the right, resulting in a shift of the spool to the right.

If a pressure of P is applied to the right-hand end and 0.5P is applied to the left-hand end, equal and opposite forces of $P \times A$ result and the valve spool is stationary.

With a pressure of P on the right-hand end and a pressure less than 0.5P on the left-hand end, net force is to the left and the valve spool moves in that direction.

The pilot valve can thus move the main spool in either direction, in a controlled manner, by varying pressure at left-hand end of the main spool from zero to full pilot pressure.

The mechanical linkage between main spool and pilot sleeve controls the flow of fluid between pilot valve and main valve, and hence controls pressure at the left-hand end of the main spool. Suppose the electrical control signal causes the pilot spool to shift left. This increases the pressure causing the main valve to shift right which in turn pushes the sleeve left. The main valve stops moving when the hole in the pilot sleeve exactly aligns with the land on the pilot spool. A change in electrical signal moving the pilot spool to the right reduces pressure at the left-hand end of the main spool by bleeding fluid back to the tank. This causes the main spool to move left until, again, pilot sleeve and pilot lands are aligned. The main valve spool thus follows the pilot spool with equal, but opposite movements.

Figure 4.41 illustrates the construction of a different type of servo valve, called a jet pipe servo. Pilot pressure is applied to a jet pipe which, with a 50% control signal, directs an equal flow into two pilot lines. A change of control signal diverts the jet flow giving unequal flows and hence unequal pressures at ends of the main spool. The main spool is linked mechanically to the jet pipe, causing it to move to counteract the applied electrical signal. Spool movement ceases when the jet pipe is again centrally located over the two pilot pipes. This occurs when the main valve spool movement exactly balances the electrical control signal.

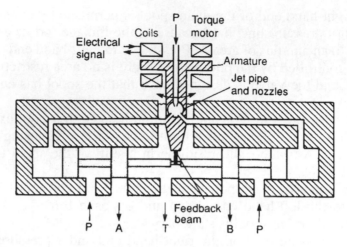

Figure 4.41   *Jet pipe servo valve*

The servo valve in Figure 4.42 is called a flapper servo and is really the inverse of the jet pipe servo. Here, pilot pressure is applied to both ends of the main spool and linked by orifices to small jets playing to a flapper which can be moved by the electrical control signal. Pressure at each end of the main spool (and hence spool movement) is determined by the flow out of each jet which, in turn, is determined by flapper position and electrical control signal.

Servo valves are generally used as part of an external control loop in a feedback control system. The principle of a feedback

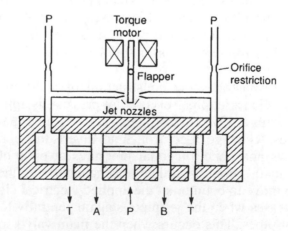

Figure 4.42   *Flapper jet servo valve*

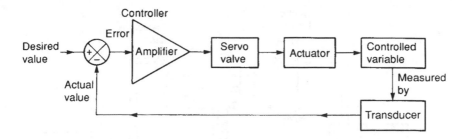

Figure 4.43 *A feedback control system*

control system is shown in Figure 4.43 where some plant variable (velocity or position, for example) is to be controlled. The plant variable is measured by a suitable transducer, and electronically compared with the desired value to give an error signal. This is amplified and used as the control signal for the servo valve.

It can be appreciated that, with small movements of the pilot spool (in Figure 4.40) and the fine jets (in Figures 4.41 and 4.42), servo valves are particularly vulnerable to dirt. Cleanliness is important in *all* aspects of pneumatics and hydraulics, but is overwhelmingly important with servo valves. A filtration level of 10 μm is normally recommended (compared with a normal filtration of 25 μm for finite position valve systems).

Servo valves which are stationary for the majority of time can stick in position due to build-up of scum around the spool. This is known, aptly, as stiction. A side effect of stiction can be a deadband where a large change of control signal is needed before the valve responds at all.

Figure 4.44 shows a purely mechanical servo used as a mechanical booster to allow a large load to be moved with minimal effort. The pilot valve body is connected to the load, and directs fluid to the fixed main cylinder. The cylinder, and hence the load, moves until pilot spool and cylinder are again aligned. Variations on the system in Figure 4.44 are used for power steering in motor cars.

Modular valves and manifolds

Valves are normally mounted onto a valve skid with piping at the rear, or underneath, to allow quick changes to be made for maintenance purposes. Piping can, however, be dispensed with almost totally by mounting valves onto a manifold block- with intercon-

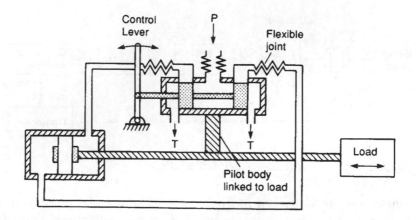

Figure 4.44 *Power assistance using mechanical servo valve*

nections formed by drilled passages in a solid block or by cut-outs on a plate-formed manifold.

Modular valve assemblies allow piping to be reduced still further. These follow standards laid down by the *Comité European des Transmission Oleophydrauliques et Pneumatiques* and are consequently known as *CETOP* modular valves.

Modular valves consist of a base plate, shown in Figure 4.45a, and a wide variety of modules which may be stacked up on top. Figures 4.45b to d show some modules available. At the top of the stack a spool valve or crossover plate is fitted. Quite complicated assemblies can be built up with minimal piping and the ease of a child's building block model.

Cartridge logic valves

These are simple two position Open/Shut valves using a poppet and seat. Figure 4.46 shows the construction and symbol for a normally open (pilot to close) valve. A normally closed (pilot to open) valve can be constructed as Figure 4.47.

Because a cartridge valve is a two position valve, four valves are needed to provide directional control. Figure 4.48 shows a typical circuit for moving a cylinder. Note these are operated in pairs by a solenoid operated two position valve; 2 and 4 cause the cylinder to extend and 1 and 3 cause the cylinder to retract. As drawn the cylinder will drive to a fully extended or fully retracted position. If the cylinder was required to hold an intermediate position the single

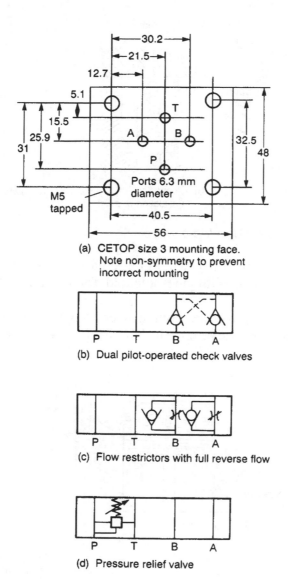

(a) CETOP size 3 mounting face.
Note non-symmetry to prevent
incorrect mounting

(b) Dual pilot-operated check valves

(c) Flow restrictors with full reverse flow

(d) Pressure relief valve

Figure 4.45 *CETOP modular valves. Examples shown are only
a small proportion of those available*

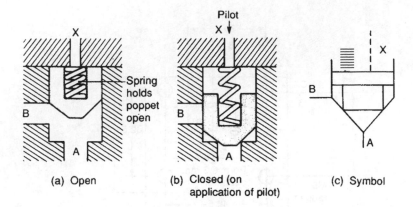

(a) Open (b) Closed (on application of pilot) (c) Symbol

Figure 4.46 *Cartridge logic valve*

two position solenoid valve would be replaced by a three position centre blocked valve with one solenoid for extend and one for retract.

At first sight this may be thought over complex compared with the equivalent spool valve circuit, but cartridge valves have some

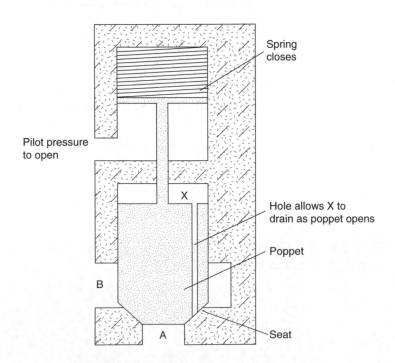

Figure 4.47 *Normally closed, pilot to open, cartridge valve*

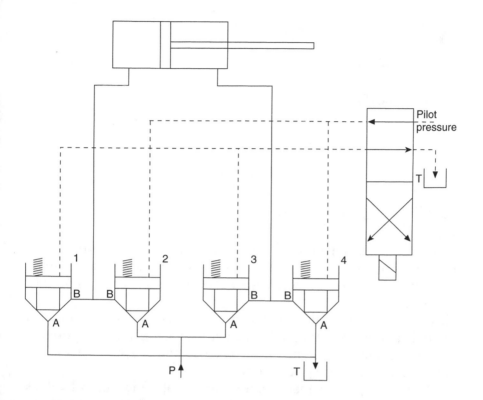

Figure 4.48 *Direction control using four cartridge valves. As shown the cyclinder will fully extend or fully retract. If two solenoid valves are used, one for open, one for close, the cylinder can hold position*

distinct advantages. Because of their construction they have very low leakage and can handle higher flows than spool valves of a similar size. They are also modular and are connected by screwing into a pre-drilled manifold. This provides high reliability and easy fault diagnosis and replacement. They are commonly used on mobile plant and with water based fluids where leakage can be a problem.

5

Actuators

A hydraulic or pneumatic system is generally concerned with moving, gripping or applying force to an object. Devices which actually achieve this objective are called actuators, and can be split into three basic types.

Linear actuators, as the name implies, are used to move an object or apply a force in a straight line. Rotary actuators are the hydraulic and pneumatic equivalent of an electric motor. This chapter discusses linear and rotary actuators.

The third type of actuator is used to operate flow control valves for process control of gases, liquids or steam. These actuators are generally pneumatically operated and are discussed with process control pneumatics in Chapter 7.

Linear actuators

The basic linear actuator is the cylinder, or ram, shown in schematic form in Figure 5.1. Practical constructional details are discussed later. The cylinder in Figure 5.1 consists of a piston, radius R, moving in a bore. The piston is connected to a rod of radius r which drives the load. Obviously if pressure is applied to port X (with port Y venting) the piston extends. Similarly, if pressure is applied to port Y (with port Z venting), the piston retracts.

The force applied by a piston depends on both the area and the applied pressure. For the extend stroke, area A is given by πR^2. For a pressure P applied to port X, the extend force available is:

$$F_c = P \, \pi \, R^2. \tag{5.1}$$

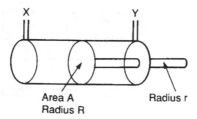

X   Y

Area A   Radius r
Radius R

Figure 5.1   *A simple cylinder*

The units of expression 5.1 depend on the system being used. If SI units are used, the force is in newtons.

Expression 5.1 gives the maximum achievable force obtained with the cylinder in a stalled condition. One example of this occurs where an object is to be gripped or shaped.

In Figure 5.2 an object of mass M is lifted at constant speed. Because the object is not accelerating, the upward force is equal to Mg newtons (in SI units) which from expression 5.1 gives the pressure in the cylinder. This is lower than the maximum system pressure; the pressure drop occurring across flow control valves and system piping. Dynamics of systems similar to this are discussed later.

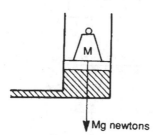

M

Mg newtons

Figure 5.2   *A mass supported by a cylinder*

When pressure is applied to port Y, the piston retracts. Total piston area here is reduced because of the rod, giving an annulus of area A_a where:

$$A_a = A - \pi r^2$$

and r is the radius of the rod. The maximum retract force is thus:

$$F_r = P A_a = P(A - \pi r^2). \tag{5.2}$$

This is lower than the maximum extend force. In Figure 5.3 identical pressure is applied to both sides of a piston. This produces an

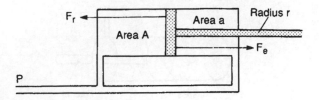

Figure 5.3   *Pressure applied to both sides of piston*

extend force F_c given by expression 5.1, and a retract force F_r given by expression 5.2. Because F_c is greater than F_r, the cylinder extends.

Normally the ratio A/A_a is about 6/5. In the cylinder shown in Figure 5.4, the ratio A/A_a of 2:1 is given by a large diameter rod. This can be used to give an equal extend and retract force when connected as shown. (The servo valve of Figure 4.40 also uses this principle.)

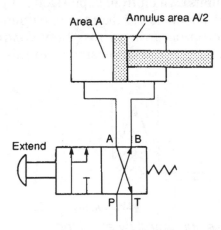

Figure 5.4   *Cylinder with equal extend/retract force*

Cylinders shown so far are known as double-acting, because fluid pressure is used to extend and retract the piston. In some applications a high extend force is required (to clamp or form an object) but the retract force is minimal. In these cases a single-acting cylinder (Figure 5.5) can be used, which is extended by fluid but retracted by a spring. If a cylinder is used to lift a load, the load itself can retract the piston.

Single-acting cylinders are simple to drive (particularly for pneumatic cylinders with quick exhaust valves (see Chapter 4)) but the

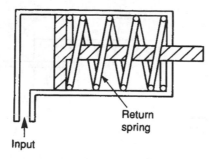

Figure 5.5   *Single-acting cylinder*

extend force is reduced and, for spring-return cylinders, the Figure length of the cylinder is increased for a given stroke to accommodate the spring.

A double rod cylinder is shown in Figure 5.6a. This has equal fluid areas on both sides of the piston, and hence can give equal forces in both directions. If connected as shown in Figure 5.3 the piston does not move (but it can be shifted by an outside force). Double rod cylinders are commonly used in applications similar to Figure 5.6b where a dog is moved by a double rod cylinder acting via a chain.

The speed of a cylinder is determined by volume of fluid delivered to it. In the cylinder in Figure 5.7 the piston, of area A, has moved a distance d. This has required a volume V of fluid where:

$$V = A \, d \qquad\qquad (5.3)$$

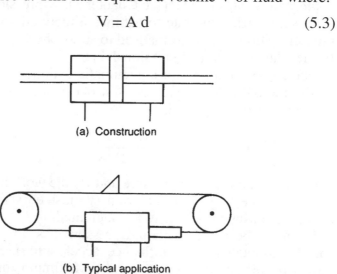

Figure 5.6   *Double rod cylinder (with equal extend/retract force)*

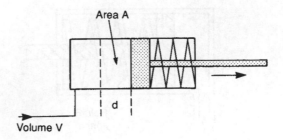

Figure 5.7   *Derivation of cylinder speed*

If the piston moves at speed v, it moves distance d in time t where:

$$t = d/v$$

Flow rate, V_f, to achieve speed v is thus:

$$V_f = \frac{A\,d}{t}$$
$$= A\,v \qquad\qquad (5.4)$$

The flow rate units of expression 5.4 depend on the units being used. If d is in metres, v in metres min^{-1} and A in metres^2, flow rate is in metres^3 min^{-1}.

In pneumatic systems, it should be remembered, it is normal to express flow rates in STP (see Chapter 3). Expression 5.4 gives the fluid volumetric flow rate to achieve a required speed at working pressure. This must be normalised to atmospheric pressure by using Boyle's law (given in expression 1.17).

The air consumption for a pneumatic cylinder must also be normalised to STP. For a cylinder of stroke S and piston area A, normalised air consumption is:

$$\text{Volume/stroke} = S\,A\,\frac{(P_a + P_w)}{P_a} \qquad\qquad (5.5)$$

where P_a is atmospheric pressure and P_w the working pressure. The repetition rate (e.g. 5 strokes min^{-1}) must be specified to allow mean air consumption rate to be calculated.

It should be noted that fluid pressure has no effect on piston speed (although it does influence acceleration). Speed is determined by piston area and flow rate. Maximum force available is unrelated to flow rate, instead being determined by line pressure and piston area. Doubling the piston area while keeping flow rate and line pressure constant, for example, gives half speed but doubles the maximum force. Ways in which flow rate can be controlled are discussed later.

Construction

Pneumatic and hydraulic linear actuators are constructed in a similar manner, the major differences arising out of differences in operating pressure (typically 100 bar for hydraulics and 10 bar for pneumatics, but there are considerable deviations from these values).

Figure 5.8 shows the construction of a double-acting cylinder. Five locations can be seen where seals are required to prevent leakage. To some extent, the art of cylinder design is in choice of seals, a topic discussed further in a later section.

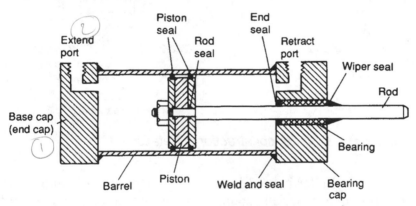

Figure 5.8   *Construction of a typical cylinder*

There are five basic parts in a cylinder; two end caps (a base cap and a bearing cap) with port connections, a cylinder barrel, a piston and the rod itself. This basic construction allows fairly simple manufacture as end caps and pistons are common to cylinders of the same diameter, and only (relatively) cheap barrels and rods need to be changed to give different length cylinders. End caps can be secured to the barrel by welding, tie rods or by threaded connection. Basic constructional details are shown in Figure 5.9.

The inner surface of the barrel needs to be very smooth to prevent wear and leakage. Generally a seamless drawn steel tube is used which is machined (honed) to an accurate finish. In applications

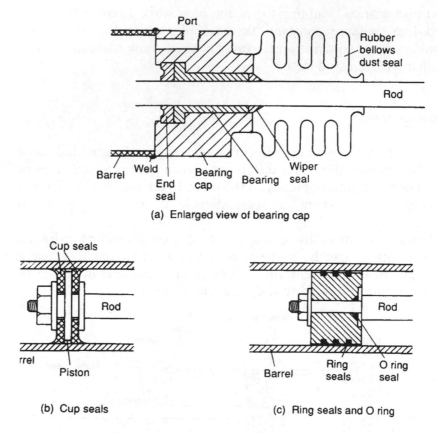

(a) Enlarged view of bearing cap

(b) Cup seals

(c) Ring seals and O ring

Figure 5.9 *Cylinder constructional details*

where the cylinder is used infrequently or may come into contact with corrosive materials, stainless steel, aluminium or brass tube may be used.

Pistons are usually made of cast iron or steel. The piston not only transmits force to the rod, but must also act as a sliding bearing in the barrel (possibly with side forces if the rod is subject to a lateral force) and provide a seal between high and low pressure sides. Piston seals are generally used between piston and barrel. Occasionally small leakage can be tolerated and seals are not used. A bearing surface (such as bronze) is deposited on to the piston surface then honed to a finish similar to that of the barrel.

The surface of the cylinder rod is exposed to the atmosphere when extended, and hence liable to suffer from the effects of dirt, moisture and corrosion. When retracted, these antisocial materials may be drawn back inside the barrel to cause problems inside the cylinder. Heat treated chromium alloy steel is generally used for strength and to reduce effects of corrosion.

A wiper or scraper seal is fitted to the end cap where the rod enters the cylinder to remove dust particles. In very dusty atmospheres external rubber bellows may also be used to exclude dust (Figure 5.9a) but these are vulnerable to puncture and splitting and need regular inspection. The bearing surface, usually bronze, is fitted behind the wiper seal.

An internal sealing ring is fitted behind the bearing to prevent high pressure fluid leaking out along the rod. The wiper seal, bearing and sealing ring are sometimes combined as a cartridge assembly to simplify maintenance. The rod is generally attached to the piston via a threaded end as shown in Figures 5.9b and c. Leakage can occur around the rod, so seals are again needed. These can be cap seals (as in Figure 5.9b) which combine the roles of piston and rod seal, or a static O ring around the rod (as in Figure 5.9c).

End caps are generally cast (from iron or aluminium) and incorporate threaded entries for ports. End caps have to withstand shock loads at extremes of piston travel. These loads arise not only from fluid pressure, but also from kinetic energy of the moving parts of the cylinder and load.

These end of travel shock loads can be reduced with cushion valves built into the end caps. In the cylinder shown in Figure 5.10, for example, exhaust fluid flow is unrestricted until the plunger

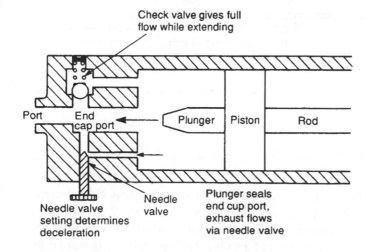

Figure 5.10 *Cylinder cushioning*

enters the cap. The exhaust flow route is now via the deceleration valve which reduces the speed and the end of travel impact. The deceleration valve is adjustable to allow the deceleration rate to be set. A check valve is also included in the end cap to bypass the deceleration valve and give near full flow as the cylinder extends. Cushioning in Figure 5.10 is shown in the base cap, but obviously a similar arrangement can be incorporated in bearing cap as well.

Cylinders are very vulnerable to side loads, particularly when fully extended. In Figure 5.11a a cylinder with a 30 cm stroke is fully extended and subject to a 5 kg side load. When extended there is typically 1 cm between piston and end bearing. Simple leverage will give side loads of 155 kg on the bearing and 150 kg on the piston seals. This magnification of side loading increases cylinder wear. The effect can be reduced by using a cylinder with a longer stroke, which is then restricted by an internal stop tube as shown in Figure 5. 11b.

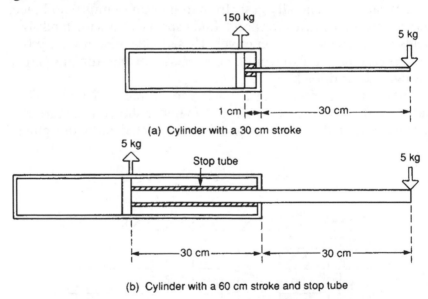

(a) Cylinder with a 30 cm stroke

(b) Cylinder with a 60 cm stroke and stop tube

Figure 5.11   *Side loads and the stop tube*

The stroke of a simple cylinder must be less than barrel length, giving at best an extended/retracted ratio of 2:1. Where space is restricted, a telescopic cylinder can be used. Figure 5.12 shows the construction of a typical double-acting unit with two pistons. To extend, fluid is applied to port A. Fluid is applied to both sides of

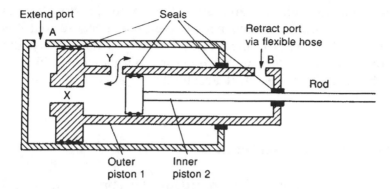

Figure 5.12 *Two-stage telescopic piston*

piston 1 via ports X and Y, but the difference in areas between sides of piston 1 causes the piston to move to the right.

To retract, fluid is applied to port B. A flexible connection is required for this port. When piston 2 is driven fully to the left, port Y is now connected to port B, applying pressure to the right-hand side of piston 1 which then retracts.

The construction of telescopic cylinders requires many seals which makes maintenance complex. They also have smaller force for a given diameter and pressure, and can only tolerate small side loads.

Pneumatic cylinders are used for metal forming, an operation requiring large forces. Pressures in pneumatic systems are lower than in hydraulic systems, but large impact loads can be obtained by accelerating a hammer to a high velocity then allowing it to strike the target.

Such devices are called impact cylinders and operate on the principle illustrated in Figure 5.13. Pressure is initially applied to port

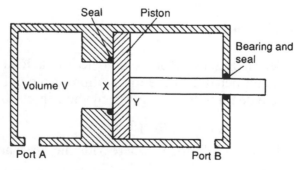

Figure 5.13 *An impact cylinder*

B to retract the cylinder. Pressure is then applied to both ports A and B, but the cylinder remains in a retracted state because area X is less than area Y. Port B is then vented rapidly. Immediately, the full piston area experiences port A pressure. With a large volume of gas stored behind the piston, it accelerates rapidly to a high velocity (typically 10m s^{-1}).

Mounting arrangements

Cylinder mounting is determined by the application. Two basic types are shown in Figure 5.14. The clamp of Figure 5.14a requires a simple fixed mounting. The pusher of Figure 5.15b requires a cylinder mount which can pivot.

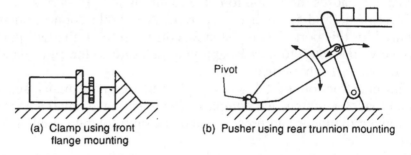

(a) Clamp using front   (b) Pusher using rear trunnion mounting
    flange mounting

Figure 5.14   *Basic mounting types*

Figure 5.15 shows various mounting methods using these two basic types. The effects of side loads should be considered on non-centreline mountings such as the foot mount. Swivel mounting obviously requires flexible pipes.

Cylinder dynamics

The cylinder in Figure 5.16a is used to lift a load of mass M. Assume it is retracted, and the top portion of the cylinder is pressurised. The extending force is given by the expression:

$$F = P_1 A - P_2 a. \tag{5.6}$$

To lift the load at all, F>Mg+f where M is the mass and f the static frictional force.

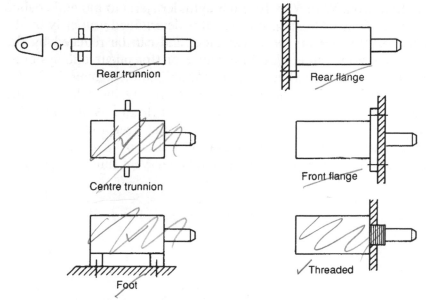

Figure 5.15 *Methods of cylinder mounting*

The response of this simple system is shown in Figure 5.16b. At time W the rod side of the cylinder is vented and pressure is applied to the other side of the piston. The pressure on both sides of the piston changes exponentially, with falling pressure P_2 changing slower than inlet pressure P_1, because of the larger volume. At time X, extension force P_1A is larger than P_2a, but movement does not start until time Y when force, given by expression 5.6, exceeds mass and frictional force.

The load now accelerates with acceleration given by Newton's law:

$$\text{acceleration} = \frac{F_a}{M} . \tag{5.7}$$

Where $F_a = P_1A - P_2a - Mg - f$.

It should be remembered that F_a is not constant, because both P_1 and P_2 will be changing. Eventually the load will reach a steady velocity, at time Z. This velocity is determined by maximum input flow rate or maximum outlet flow rate (whichever is lowest). Outlet pressure P_2 is determined by back pressure from the outlet line to tank or atmosphere, and inlet pressure is given by the expression:

$$P_1 = \frac{Mg + f + P_2a}{A}$$

The time from W to Y, before the cylinder starts to move, is called the 'dead time' or 'response time'. It is determined primarily by the decay of pressure on the outlet side, and can be reduced by depressurising the outlet side in advance or (for pneumatic systems) by the use of quick exhaust valves (described in Chapter 4).

The acceleration is determined primarily by the inlet pressure and the area of the inlet side of the piston (term P_1A in expression 5.6). The area, however, interacts with the dead time – a larger area, say, gives increased acceleration but also increases cylinder volume and hence extends the time taken to vent fluid on the outlet side.

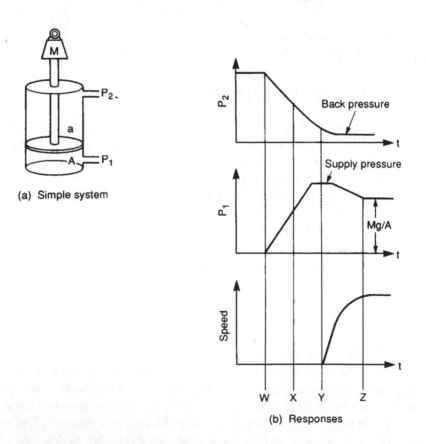

Figure 5.16   *Cylinder dynamics*

Seals

Leakage from a hydraulic or pneumatic system can be a major problem, leading to loss of efficiency, increased power usage, temperature rise, environmental damage and safety hazards.

Minor internal leakage (round the piston in a double-acting cylinder, for example) can be of little consequence and may even be deliberately introduced to provide lubrication of the moving parts.

External leakage, on the other hand, is always serious. In pneumatic systems, external leakage is noisy; with hydraulic systems, external loss of oil is expensive as lost oil has to be replaced, and the resulting pools of oil are dangerous and unsightly.

Mechanical components (such as pistons and cylinders) cannot be manufactured to sufficiently tight tolerances to prevent leakage (and even if they could, the resultant friction would be unacceptably high). Seals are therefore used to prevent leakage (or allow a controlled leakage). To a large extent, the art of designing an actuator is really the art of choosing the right seals.

The simplest seals are 'static seals' (Figure 5.17) used to seal between stationary parts. These are generally installed once and forgotten. A common example is the gasket shown in a typical application in Figure 5.17a. The O ring of Figure 5.17b is probably the most used static seal, and comprises a moulded synthetic ring with a round cross section when unloaded. O rings can be specified in terms of inside diameter (ID) for fitting onto shafts, or outside diameter (OD) for fitting into bores.

When installed, an O ring is compressed in one direction. Application of pressure causes the ring to be compressed at right angles, to give a positive seal against two annular surfaces and one flat surface. O rings give effective sealing at very high pressures.

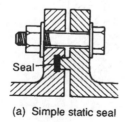

(a) Simple static seal

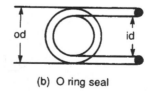

(b) O ring seal

Figure 5.17   *Static seals*

O rings are primarily used as static seals because any movement will cause the seal to rotate allowing leakage to occur.

Where a seal has to be provided between moving surfaces, a dynamic seal is required. A typical example is the end or cup seal shown, earlier, in Figure 5.9a. Pressure in the cylinder holds the lip of the seal against the barrel to give zero leakage (called a 'positive seal'). Effectiveness of the seal increases with pressure, and leakage tends to be more of a problem at low pressures.

The U ring seal of Figure 5.18 works on the same principle as the cup seal. Fluid pressure forces the two lips apart to give a positive seal. Again, effectiveness of the seal is better at high pressure. Another variation on the technique is the composite seal of Figure 5.19. This is similar in construction to the U ring seal, but the space between the lips is filled by a separate ring. Application of pressure again forces the lips apart to give a positive seal.

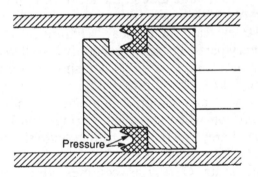

Figure 5.18 *The U ring seal*

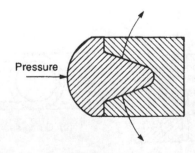

Figure 5.19 *The composite seal*

At high pressures there is a tendency for a dynamic seal to creep into the radial gap, as shown in Figure 5.20a leading to trapping of the seal and rapid wear. This can be avoided by the inclusion of an anti-extrusion ring behind the seal, as in Figure 5.20b.

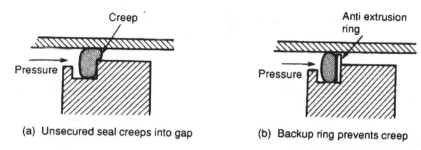

(a) Unsecured seal creeps into gap   (b) Backup ring prevents creep

Figure 5.20   *Anti-extrusion ring*

Seals are manufactured from a variety of materials, the choice being determined by the fluid, its operating pressure and the likely temperature range. The earliest material was leather and, to a lesser extent, cork but these have been largely superseded by plastic and synthetic rubber materials. Natural rubber cannot be used in hydraulic systems as it tends to swell and perish in the presence of oil.

The earliest synthetic seal material was neoprene, but this has a limited temperature range (below 65°C). The most common present-day material is nitrile (buna-N) which has a wider temperature range (–50°C to 100°C) and is currently the cheapest seal material. Silicon has the highest temperature range (–100°C to +250°C) but is expensive and tends to tear.

In pneumatic systems viton (–20°C to 190°C) and teflon (–80°C to +200°C) are the most common materials. These are more rigid and are often used as wiper or scraper seals on cylinders.

Synthetic seals cannot be used in applications where a piston passes over a port orifice which nicks the seal edges. Here metallic ring seals must be used, often with the rings sitting on O rings, as illustrated in Figure 5.21.

Seals are delicate and must be installed with care. Dirt on shafts or barrels can easily nick a seal as it is slid into place. Such damage may not be visible to the eye but can cause serious leaks. Sharp edges can cause similar damage so it is usual for shaft ends and groove edges to be chamfered.

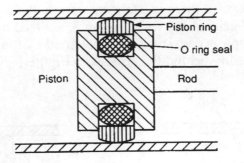

Figure 5.21 *Combined piston ring and O ring seal (not to scale)*

Rotary actuators

Rotary actuators are the hydraulic or pneumatic equivalents of electric motors. For a given torque, or power, a rotary actuator is more compact than an equivalent motor, cannot be damaged by an indefinite stall and can safely be used in an explosive atmosphere. For variable speed applications, the complexity and maintenance requirements of a rotary actuator are similar to a thyristor-controlled DC drive, but for fixed speed applications, the AC induction motor (which can, for practical purposes, be fitted and forgotten) is simpler to install and maintain.

A rotary actuator (or, for that matter, an electric motor) can be defined in terms of the torque it produces and its running speed, usually given in revs per minute (rpm). Definition of torque is illustrated in Figure 5.22, where a rotary motion is produced against a force of F newtons acting at a radial distance d metres from a shaft centre. The device is then producing a torque T given by the expression:

$$T = Fd \text{ Nm.} \qquad (5.8)$$

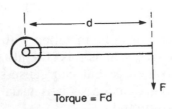

Torque = Fd

Figure 5.22 *Definition of torque*

In Imperial units, F is given in pounds force, and d in inches or feet to give T in lbf ins or lbf ft. It follows that 1 Nm = 8.85 lbf ins.

The torque of a rotary actuator can be specified in three ways. Starting torque is the torque available to move a load from rest. Stall torque must be applied by the load to bring a running actuator to rest, and running torque is the torque available at any given speed. Running torque falls with increasing speed, typical examples being shown on Figure 5.23. Obviously, torque is dependent on the applied pressure; increasing the pressure results in increased torque, as shown.

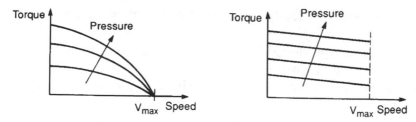

Figure 5.23 *Torque/speed curves for rotary actuators*

The output power of an actuator is related to torque and rotational speed, and is given by the expression:

$$P = \frac{T\,R}{9550}\,kw. \qquad (5.9)$$

where T is the torque in newton metre and R is the speed in rpm. In Imperial units the expression is:

$$P = \frac{T\,R}{5252}\,hp. \qquad (5.10)$$

where T is in lbsf ft (and R is in rpm) or:

$$P = \frac{T\,R}{63024}\,hp. \qquad (5.11)$$

where T is in lbsf ins.

Figure 5.23 illustrates how running torque falls with increasing speed, so the relationship between power and speed has the form of Figure 5.24, with maximum power at some (defined) speed. Power like the torque, is dependent on applied pressure.

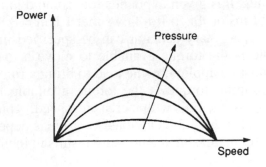

Figure 5.24   *Power/speed curve for pneumatic rotary actuator*

The torque produced by a rotary actuator is directly related to fluid pressure; increasing pressure increases maximum available torque. Actuators are often specified by their torque rating, which is defined as:

$$\text{torque rating} = \frac{\text{torque}}{\text{pressure}}$$

In Imperial units a pressure of 100 psi is used, and torque is generally given in lbf ins.

The allowable pressure for an actuator is defined in terms of pressure rating (maximum applicable pressure without risk of permanent damage), and pressure range (the maximum and minimum pressures between which actuator performance is defined).

Fluid passes through an actuator as it rotates. For hydraulic actuators, displacement is defined as the volume of fluid used for one motor rotation. For a given design of motor, available torque is directly proportional to displacement. For pneumatic actuators, the air usage per revolution at a specified pressure is generally given in terms of STP (see Chapter 3).

Rotational speed is given by the expression:

$$\text{rotational speed} = \frac{\text{fluid flow rate}}{\text{displacement}}$$

With the torque rate and displacement fixed for a chosen motor, the user can control maximum available torque and speed by adjusting, respectively, pressure setting and flow rate of fluid to the actuator.

Constructional details

In electrical systems, there are many similarities between electrical generators and electric motors. A DC generator, for example, can be run as a motor. Similarly, a DC motor can be used as a generator. Similar relationships exist between hydraulic pumps and motors and between pneumatic compressors and motors. This similarity is extended as manufacturers use common parts in pumps, compressors and motors to simplify users' spares holdings.

The similarity between pumps, compressors and motors extends to graphic symbols. The schematic symbols of Figure 5.25 are used to show hydraulic and pneumatic motors. Internal leakage always occurs in a hydraulic motor, and a drain line, shown dotted, is used to return the leakage fluid to the tank. If this leakage return is inhibited the motor may pressure lock and cease to rotate or even suffer damage.

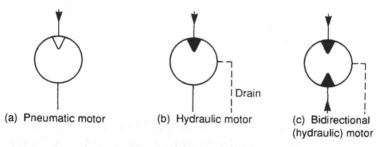

(a) Pneumatic motor   (b) Hydraulic motor   (c) Bidirectional (hydraulic) motor

Figure 5.25 *Rotary actuator symbols*

There are three basic designs of rotary or pump compressor; the gear pump, the vane pump and various designs of piston pump or compressor described earlier in Chapter 2. These can also be used as the basis of rotary actuators. The principles of hydraulic and pneumatic devices are very similar, but the much higher hydraulic pressures give larger available torques and powers despite lower rotational speeds.

Figure 5.26 shows the construction of a gear motor. Fluid enters at the top and pressurises the top chamber. Pressure is applied to two gear faces at X, and a single gear face at Y. There is, thus, an imbalance of forces on the gears resulting in rotation as shown. Gear motors suffer from leakage which is more pronounced at low speed. They thus tend to be used in medium speed, low torque applications.

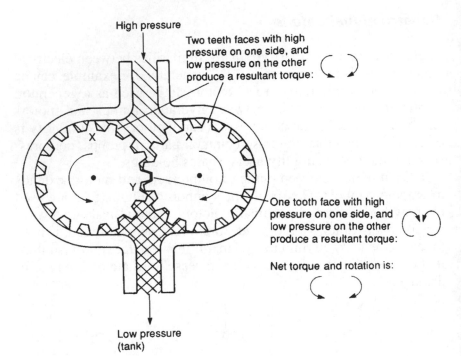

Figure 5.26   *A gear motor*

A typical vane motor construction is illustrated in Figure 5.27. It is very similar to the construction of a vane pump. Suffering from less leakage than the gear motor, it is typically used at lower speeds. Like the vane pump, side loading occurs on the shaft of a single vane motor. These forces can be balanced by using a dual design similar to the pump shown in Figure 2.10b. In a vane pump, vanes are held out by the rotational speed. In a vane motor, however, rotational speed is probably quite low and the vanes are held out, instead, by fluid pressure. An in-line check valve can be used, as in Figure 5.28, to generate a pressure which is always slightly higher than motor pressure.

Piston motors are generally most efficient and give highest torques, speeds and powers. They can be of radial design similar to the pump of Figures 2.12 and 2.13, or in-line (axial) design similar to those of Figures 2.14 and 2.15. Radial piston motors tend to be most common in pneumatic applications, with in-line piston motors most common in hydraulics. The speed of the piston motor can be varied by adjusting the angle of the swash plate (in a similar manner to which delivery volume of an in-line piston pump can be varied).

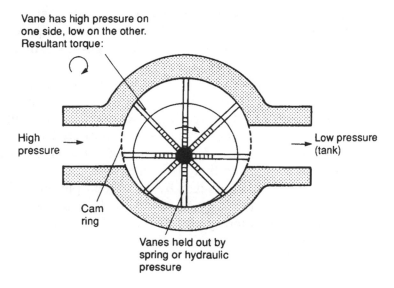

Figure 5.27 *A vane motor*

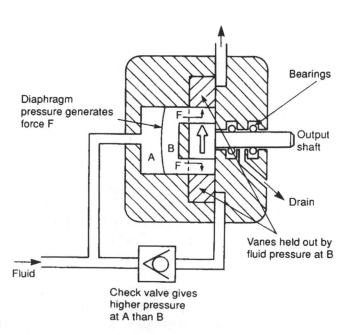

Figure 5.28 *Vane operation in hydraulic motor*

Turbine-based motors can also be used in pneumatics where very high speeds (up to 500,000 rpm) but low torques are required. A common application of these devices is the high-speed dentist's drill.

All the rotary actuators described so far have been pneumatic or hydraulic equivalents of electric motors. However, rotary actuators with a limited travel (say 270°) are often needed to actuate dampers or control large valves. Some examples are illustrated in Figure 5.29. The actuator in Figure 5.29a is driven by a single vane coupled to the output shaft. In that of Figure 5.29b, a double-acting piston is coupled to the output shaft by a rack and pinion. In both cases the shaft angle can be finely controlled by fluid applied to the ports. These have the graphic symbol shown in Figure 5.29c.

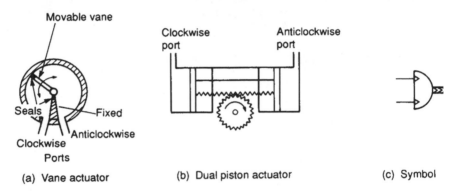

(a) Vane actuator   (b) Dual piston actuator   (c) Symbol

Figure 5.29   *Limited motion rotary actuators*

Application notes

Speed control

The operational speed of an actuator is determined by the fluid flow rate and the actuator area (for a cylinder) or the displacement (for a motor). The physical dimensions are generally fixed for an actuator, so speed is controlled by adjusting the fluid flow to (or restricting flow from) the actuator. Rotary actuator speed can also be controlled by altering swash plate angle.

The compressibility of air, normally advantageous where smooth operation is concerned, makes flow control more difficult for pneumatic than hydraulic systems. Although techniques described below can be applied in pneumatics, precise slow-speed control of a pneumatic actuator is achieved with external devices described later.

There are essentially four ways in which fluid flow can be controlled. The first is shown in Figure 5.30, where a pump delivers a fluid volume V per minute. Because the pump is a fixed displacement device this volume of fluid *must* go either back to the tank or to the actuator. When the control valve moves from its centre position, the actuator moves with a velocity:

$$v = \frac{V}{A}$$

where A is the piston area. If pump delivery volume V can be adjusted (by altering swash plate angle, say,) *and* the pump feeds no other device, no further speed control is needed.

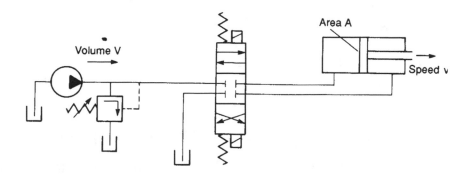

Figure 5.30 *Speed control by pump volume*

Most systems, however, are not that simple. In the second speed control method of Figure 5.31, a pump controls many devices and is loaded by a solenoid-operated valve (see Chapter 2). Unused fluid goes back to the tank via relief valve V_3. The pump output is higher than needed by any individual actuator, so a flow restrictor is used to set the flow to each actuator. This is known as a 'meter in' circuit, and is used where a force is needed to move a load. Check valve V_1 gives a full-speed retraction, and check valve V_2 provides a small back pressure to avoid the load running away. The full pump delivery is produced when the pressure reaches the setting of relief valve V_3, so there is a waste of energy and unnecessary production of heat in the fluid.

If the load can run away from the actuator, the third speed control method; the 'meter out' circuit of Figure 5.32 must be used. As

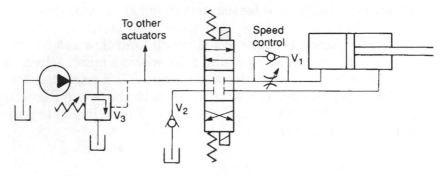

Figure 5.31   *Meter in speed control*

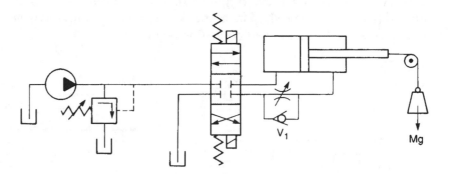

Figure 5.32   *Meter out speed control for overhauling load*

drawn, this again gives a controlled extension speed, and full retraction speed (allowed by check valve V_1). As before, the pump delivers fluid at a pressure set by the relief valve, leading to heat generation.

Finally, in the fourth speed control method of Figure 5.33, a bleed-off valve V_1 is incorporated. This returns a volume v back to

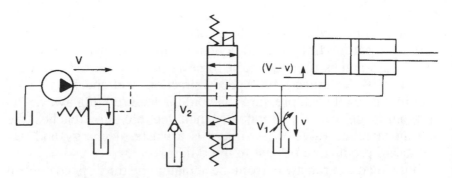

Figure 5.33   *Bleed-off speed control*

the tank, leaving a volume V-v to go to the actuator (where V is the pump delivery volume). Pump pressure is now determined by the required actuator pressure, which is lower than the pressure set on the relief valve. The energy used by the pump is lower, and less heat is generated. The circuit can, however, only be used with a load which opposes motion. Check valve V_2 again gives a small back pressure.

Any unused fluid from the pump is returned to the tank at high pressure leading to wasted energy; even with the more efficient 'bleed'-off circuit. One moral, therefore, is to have a pump delivery volume no larger than necessary.

Figures 5.31 to 5.33 imply flow, and hence speed, is set by a simple restriction in piping to the actuator. While a simple restriction reduces flow and allows speed to be reduced, in practice a true flow control valve is needed which delivers a fixed flow regardless of line pressure or fluid temperature.

An ideal flow controller operates by maintaining a constant pressure drop across an orifice restriction in the line, the rate being adjusted by altering orifice size. The construction of such a device is shown in Figure 5.34. The orifice is formed by a notch in a shaft which can be rotated to set the flow. The pressure drop across the orifice is the difference in pressure between points X and Y, and is applied to the moveable land. The pressure at X, in conjunction with the spring pressure, causes a downward force, while pressure at Y causes an upward force. If the land moves up the flow reduces,

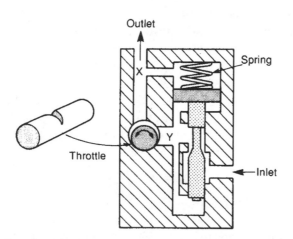

Figure 5.34   *Pressure compensated flow control valve*

if the land moves down the flow increases. The piston thus moves up and down until the pressure differential between X and Y matches the spring compressive force. The device thus maintains a constant pressure drop across the orifice, which implies constant flow through the valve, and is known as a pressure-compensated flow control valve.

Flow control valves can also be adversely affected by temperature changes which alter the viscosity of the oil. For this reason more complex flow control valves often have temperature compensation. Symbols for various types of flow control valves are given in Figure 5.35.

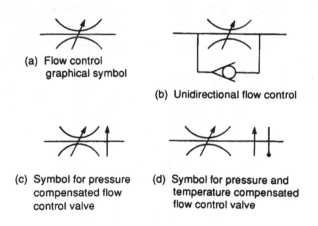

(a) Flow control
    graphical symbol

(b) Unidirectional flow control

(c) Symbol for pressure
    compensated flow
    control valve

(d) Symbol for pressure and
    temperature compensated
    flow control valve

Figure 5.35   *Flow control valves*

Discussions in this section have, so far, been concerned with hydraulic systems as compressibility of air makes speed control of pneumatic actuators somewhat crude. If a pneumatic actuator is required to act at a slow controlled speed an external *hydraulic* damper can be used, as shown in Figure 5.36. Oil is forced from one side of the hydraulic piston to the other via an adjustable flow control valve. Speeds as low as a few millimetres a minute can be accurately controlled in this manner, although the technique is physically rather cumbersome.

Actuator synchronisation

Figure 5.37 illustrates a common problem in which an unbalanced load is to be lifted by two cylinders The right-hand cylinder is

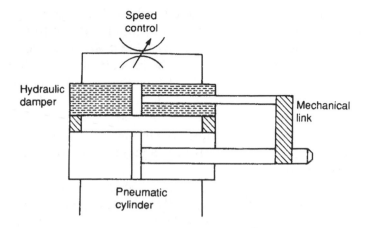

Figure 5.36 *Speed control of pneumatic cylinder*

subject to a large force F, the left-hand cylinder to a smaller force f. The right-hand piston requires a pressure of F/A to lift, while the left-hand piston needs f/A. When lift is called for on valve V_1, the pressure rises to the lower pressure f/A, and only the left-hand piston moves. The unbalanced load results in faulty operation. A similar result can occur where two, or more, cylinders operate against ill-defined frictional forces.

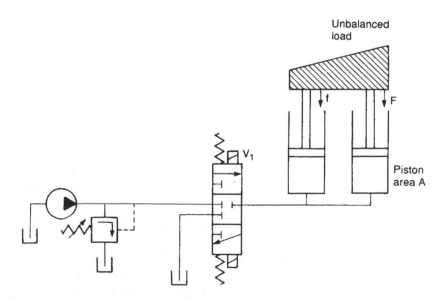

Figure 5.37 *Linked cylinders with unbalanced load*

One simple solution is the inclusion of flow regulating valves. A flow control valve can set, and hold, fluid flow to within about ±5% of nominal value, resulting in a possible positional error of 10% of the stroke. This may, or may not, be acceptable, and in the example of Figure 5.37 the cylinders would, in any case, align themselves at each end of the stroke. (When the most lightly laden and hence fastest travelling piston reaches the end of its stroke, the system pressure will rise.) This solution is not acceptable if good positional accuracy is required or rotary actuators without end stops are being driven.

The flow divider valve of Figure 5.38 works on a similar principle, dividing the inlet flow equally (to a few percent) between two outlet ports. The spool moves to maintain equal pressure drops across orifices X and Y, and hence equal flow through them.

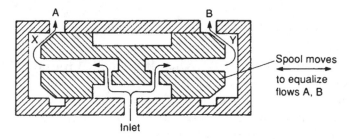

Figure 5.38   *Flow divider valve*

The displacement of a hydraulic or pneumatic motor can be accurately specified, and this forms the basis of an alternative flow divider circuit of Figure 5.39. Here fluid for two cylinders passes through two mechanically coupled motors. The mechanical coupling ensures the two motors rotate at the same speed, and hence equal flow is passed into each cylinder.

The two cylinders in Figure 5.40 are effectively in series with fluid from the annulus side of cylinder 1 going to the full bore side of cylinder 2. The cylinders are chosen, however, so that full bore area of cylinder 2 equals the annulus area of cylinder 1. Upon cylinder extension, fluid exits from cylinder 1 and causes cylinder 2 to extend. The two cylinders move at equal speed because of the equal areas.

There is, though, an unfortunate side effect. Pressure P_2 in cylinder 2 is F/a. Fluid on the full bore side of cylinder 1 has to lift the piston against force f plus the force from P_2 acting on the annulus side of the piston. Pressure P_1 is (F+f)/A; higher than would be

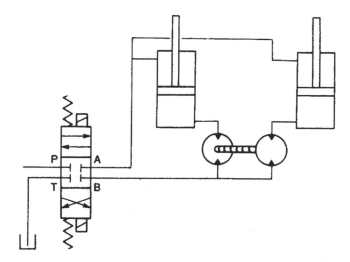

Figure 5.39 *Cylinder synchronisation with linked hydraulic motors*

required by two independent cylinders acting in parallel. The rotational speed of motors with equal displacement can similarly be synchronised by connecting them in series. Inlet pressure of the first motor is again, however, higher than needed to drive the two motors separately or in parallel.

None of these methods gives absolute synchronisation, and if actuators do not self-align at the ends of travel, some method of driving actuators individually should be included to allow intermittent manual alignment. The best solution, however, is usually to include some form of mechanical tie to ensure actuators experience equal loads and cannot get out of alignment.

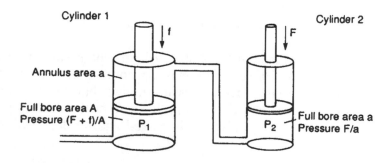

Figure 5.40 *Cylinder synchronisation with series connection*

Regeneration

A conventional cylinder can exert a larger force extending than
retracting because of the area difference between full bore and
annulus sides of the piston. The system in Figure 5.41 employs a
cylinder with a full bore/annulus ratio of 2:1, and is known as a differential cylinder.

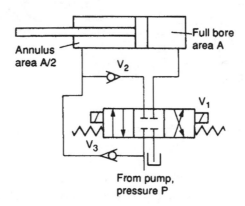

Figure 5.41   *Regeneration circuit*

Upon cylinder extension, line pressure P is applied to the right-
hand side of the piston giving a force of $P \times A$, while the left-hand
side of the piston returns oil via valve V_3 against line pressure P
producing a counter force $P \times A/2$. There is thus a net force of
$P \times A/2$ to the left. When retraction is called for, a force of $P \times A/2$
is applied to the left-hand side and fluid from the right-hand side
returns to tank at minimal pressure. Extension and retraction forces
are thus equal, at $P \times A/2$.

Counterbalance and dynamic braking

The cylinder in Figure 5.42 supports a load which can run away
when being lowered. Valve V_2, known as a counterbalance valve, is
a pressure-relief valve set for a pressure higher than F/2 (the pressure generated in the fluid on the annulus side of the piston by the
load). In the static state, valve V_2 is closed and the load holds in
place.

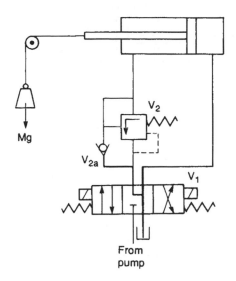

Figure 5.42 *Counterbalance circuit*

When the load is to be lowered line pressure is applied to the full bore side of the piston through valve V_1. The increased pressure causes valve V_2 to open and the load to lower. Check Valve V_2a passes fluid to raise the load.

Counterbalance valves can also be used to brake a load with high inertia. Figure 5.43 shows a system where a cylinder moves a load with high inertia. Counterbalance valves V_2 and V_3 are included in the lines to both ends of the cylinder. Cross-linked pilot lines (shown dotted as per convention) keep valve V_2 open when extending and valve V_3 open when retracting. At constant cylinder speed, therefore, valves V_2 and V_3 have little effect.

To stop the load valve V_1 is moved to its centre position, the pump unloads to tank and pilot pressure is lost, causing valves V_2 and V_3 to close. Inertia, however, maintains some cylinder movement. If, for example, the cylinder had been extending, inertia keeps it moving to the left – raising pressure on the piston's annulus side until valve V_2 reaches its pressure setting and opens. A constant deceleration force Pa (where P is the setting of valve V_1 and a is the annulus area) is applied to the load. On deceleration, fluid passes to the full bore side of the cylinder through check valve V_3a.

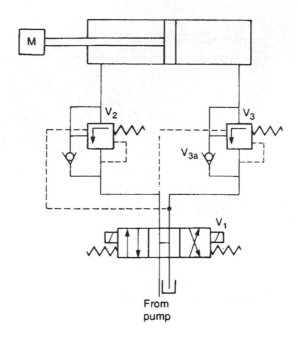

Figure 5.43   *Braking a high inertia load*

Pilot-operated check valves

Directional control valves and deceleration valves have a small, but
definite, leakage and can only be used to hold an opposing load in
position for short periods (of the order of minutes rather than hours)
without the energy wasting procedure of permanently applying
pressure to the cylinder.

A check valve can be constructed with zero leakage. The pilot-
operated check valve (described in Chapter 4) can thus be used to
'lock' an actuator in position. Figure 5.44 shows a typical example.
Valve V_2 passes fluid normally when extending, but closes when
valve V_1 is in its centre position. In this state, energy is saved by
unloading the pump to tank. Pilot line pressure opens valve V_2
when the load is to be lowered. Counterbalance valve V_3 gives a
controlled lowering but also ensures sufficient line pressure exists
on the annulus side of the cylinder to give the pilot pressure needed
to open valve V_2.

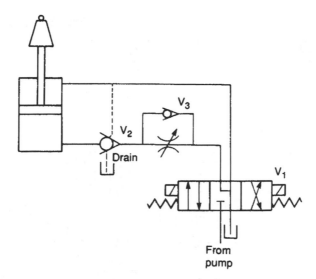

Figure 5.44 *Pilot-operated check valve used to hold an over-hauling load*

Pre-fill and compression relief

Figure 5.45 shows the hydraulic circuit for a large press. To give the required force, a large diameter cylinder is needed and, if this is driven directly, a large capacity pump is required. The circuit shown (known as a pre-fill circuit) uses a high level tank and pilot operated check valve to reduce the required pump size.

The cross-head of the press is raised and lowered by small cylinders C_1 and C_2. When valve V_1 is switched to lower, the pressure on the full bore sides of cylinders C_1 and C_2 is low and valve V_3 is closed. Valve V_4 is a counterbalance valve, giving a controlled lower. As cylinders C_1 and C_2 extend, cylinder C_3 also extends because it is mechanically coupled, drawing its fluid direct from the high level tank via pilot valve V_2.

When the cross-head contacts the load, the pressure on the full bore side of cylinders C_1 and C_2 rises. This causes valve V_3 to open, full line pressure to be applied to cylinder C_3 and check valve V_2 to close. Full operating force is now applied to the load via cylinder C_3.

When the cross-head is raised, pressure is applied to the annulus side of cylinders C_1 and C_2. This opens check valve V_2, allowing fluid in cylinder C_3 to be returned directly to tank.

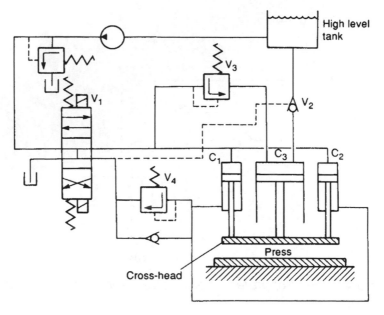

Figure 5.45   *Pre-fill circuit*

High pressure hydraulic circuits like this require care both in design and in maintenance. For most practical purposes, hydraulic fluid can be considered incompressible. In reality, it compresses by about 0.8% per 100 bar applied pressure. When high pressure and large volumes of oil are present together, sudden release of pressure can result in an explosive release of fluid. The design must, therefore, allow for the gradual release of high pressure, high volume fluid.

Large volume, high pressure valves are thus fitted with a central damping block as illustrated in Figure 5.46 to return fluid to tank slowly.

Figure 5.47 shows a common decompression circuit. When the cylinder extends, fluid passes to the full bore side via check valve V_3 as usual, with fluid pressure rising once the load is contacted. This rise in pressure keeps valve V_2 closed. When valve V_1 is returned to its centre position, the pressure decays via restriction valve RV_1. Once the pressure decays to a safe level, set by valve V_2, this valve opens allowing pressure to decay fully.

Valve V_4 is included to protect against a quick change from high pressure extend to retract, without a pause to allow the pressure to decay. When the full bore side of the cylinder is pressurised, valve V_4 is held open causing the pump to unload to tank if retract is

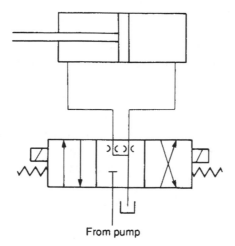

Figure 5.46   *Valve with central damping block*

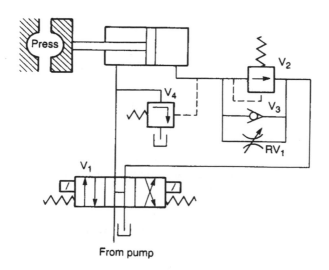

Figure 5.47   *Controlled decompression circuit*

requested before decompression is complete. Once pressure on the full bore side decays, valve V3 closes and the cylinder can retract as normal.

Bellows actuator

Many applications require a simple lift function, for example to raise a disappearing stop on a set of rollers. This function is usually provided by a pneumatic cylinder which requires space and mounting lugs. A simple alternative is the bellows of Figure 5.48. In the de-energized state the bellows are deflated and the load falls under gravity. When air is passed to the bellows they inflate lifting the load. The actuator requires minimal space in its de-energized state and is simple to mount. The only disadvantage is that the load falls under gravity and is not driven down.

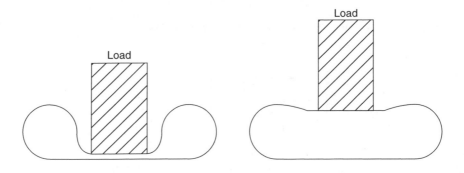

Figure 5.48   *The use of pneumatic bellows gives a simple way of raising and lowering a load*

6

Hydraulic and pneumatic accessories

Hydraulic reservoirs

A hydraulic system is closed, and the oil used is stored in a tank or reservoir to which it is returned after use. Although probably the most mundane part of the system, the design and maintenance of the reservoir is of paramount importance for reliable operation. Figure 6.1 shows details of a typical reservoir.

The volume of fluid in a tank varies according to temperature and the state of the actuators in the system, being minimum at low tem-

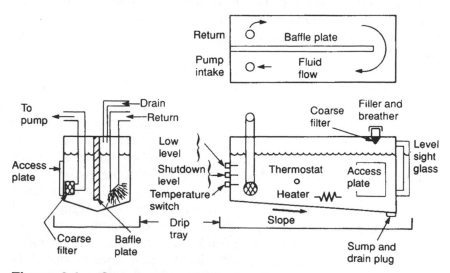

Figure 6.1 *Construction of a hydraulic reservoir*

perature with all cylinders extended, and maximum at high temperature with all cylinders retracted. Normally the tank volume is set at the larger of four times the pump draw per minute or twice the external system volume. A substantial space must be provided above the fluid surface to allow for expansion and to prevent any froth on the surface from spilling out.

The tank also serves as a heat exchanger, allowing fluid heat to be removed. To obtain maximum cooling, fluid is forced to follow the walls of the tank, from the return line to pump suction inlet, by a baffle plate down the tank centre line. This plate also encourages any contamination to fall to the tank bottom before reaching the pump inlet, and allows any entrapped air to escape to the surface. The main return line should enter from the top of the tank to preclude the need for a check valve and end below the minimum tank level to prevent air being drawn into the oil. The return flow should emerge into the tank through a diffuser with a low velocity of around 0.3 m/sec to prevent disturbance of any deposits at the base of the tank. The flow should be directed at the tank wall to assist cooling.

If, as is commonly the case, the external components outside the tank are below the oil level in the tank the return line should be equipped with a removable anti-siphonage plug. This should be removed to allow air into the return line before any external components are disconnected. Without this precaution a siphon backflow can occur which is very difficult to stop. If you have never encountered it before the sudden and apparently unstoppable flow of oil from the return pipe on disconnection can be very surprising.

Low pressure returns (such as drains from motors or valves) must be returned above fluid level to prevent back pressure and formation of hydraulic locks.

Fluid level is critical. If it is too low, a whirlpool forms above the pump inlet, resulting in air being drawn into the pump. This air results in maloperation, and will probably result in pump damage.

A level sight glass is essential to allow maintenance checks to be carried out. The only route for oil to leave a hydraulic system is, of course, by leaks so the cause of any gross loss of fluid needs investigation. In all bar the smallest and simplest systems, two electrical float switches are generally included giving a remote (low level) warning indication and a last ditch (very low level) signal which leads to automatic shutdown of the pump before damage can occur.

The temperature of fluid in the tank also needs monitoring and as an absolute minimum a simple visual thermometer should be included. The ideal temperature range is around 45 to 50°C and, usually, the problem is keeping the temperature down to this level. Ideally an electrical over-temperature switch is used to warn the user when oil temperature is too high.

When the system is used intermittently, or started up from cold, oil temperature can be too low, leading to sluggish operation and premature wear. A low temperature thermostat and electrical heater may be included to keep the oil at an optimum temperature when the system is not in use.

Reservoirs are designed to act as collecting points for all the dirt particles and contamination in the system and are generally constructed with a V-shaped cross section forming a sump. A slight slope ensures contamination collects at the lower end where a drain plug is situated. Often magnetic drain plugs are used to trap metallic particles.

Reservoirs should be drained periodically for cleaning, and a removable man access plate is included for this purpose. This is *not* the most attractive of jobs!

Oil is added through a filler cap in the tank top. This doubles as a breather allowing air into and out of the tank as the volume of fluid changes. A coarse filter below the breather prevents contamination entering the tank as fluid is added.

Tank air filters are commonly forgotten in routine maintenance. The oil in a typical tank changes considerably during operation as temperatures change and actuators operate. This change in volume is directly reflected in air changes both in and out of the tank. The only route for this air flow is through the filters. If these become blocked the tank may become pressurised and fail disastrously.

Reservoirs are generally constructed from welded steel plate with thin side walls to encourage heat loss. The inside of the tank is shot blasted then treated with protective paint to prevent formation of rust particles.

At some time in the life of a hydraulic system there will eventually be oil spillage around the tank, whether from leakage, over-enthusiastic filling or careless maintenance. It is therefore good practice to put substantial drip trays under reservoir pumps and associated valves to limit oil spread when the inevitable mishaps occur.

Hydraulic accumulators

In a simple hydraulic system, the pump size (delivery rate and hence motor power) is determined by the maximum requirements of the actuators. In Figure 6.2 a system operates intermittently at a

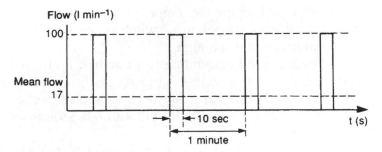

Figure 6.2   *A simple system with uneven demands. To supply this without an accumulator a 100 l min⁻¹ is required although the mean flow is only 17 1/min*

pressure of between 150 and 200 bar, needing a flow rate of 100 1 min⁻¹ for 10 s at a repetition rate of 1 minute. With a simple system (pump, pressure regulator and loading valve) this requires a 200 bar, 100 1 min⁻¹ pump (driven by about a 50 hp motor) which spends around 85% of its time unloading to tank.

In Figure 6.3a a storage device called an accumulator has been added to the system. This can store, and release, a quantity of fluid at the required system pressure. In many respects it resembles the operation of a capacitor in an electronic power supply.

The operation is shown in Figure 6.3b. At time A the system is turned on, and the pump loads causing pressure to rise as the fluid is delivered to the accumulator via the non-return valve V_3. At time B, working pressure is reached and a pressure switch on the accumulator causes the pump to unload. This state is maintained as non-return valve V_3 holds the system pressure.

The actuator operates between times C and D. This draws fluid from the accumulator causing a fall of system pressure. The pressure switch on the accumulator puts the pump on load again but it takes until time E before the accumulator is charged ready for the next actuator movement at time F.

An accumulator reduces pump requirements. The original system required a 100 1 min⁻¹ pump. With an accumulator, however, a pump only needs to provide 17 1 min⁻¹ (that is, 100 1 min⁻¹ for 10 secs every minute). Pump size, and hence motor size, have been reduced by a factor of six with obvious cost and space savings, plus gains in ancillary equipment such as motor starters and cabling. There is no gain in the energy used; with the simple system a 50 hp motor loads for 17% of the time, with an accumulator a 10 hp motor loads for about 90% of the time.

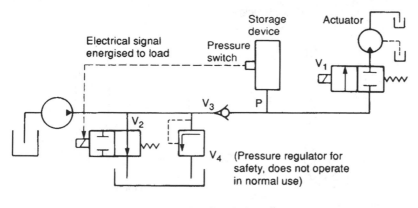

(a) Circuit diagram

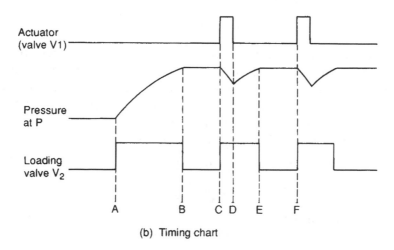

(b) Timing chart

Figure 6.3 *System with an accumulator*

Most accumulators operate by compressing a gas (although older and smaller accumulators may work by compressing a spring or lifting a weight with a cylinder). The most common form is the gas-filled bladder accumulator shown in Figure 6.4. Gas is precharged to some pressure with the accumulator empty of fluid when the whole of the accumulator is filled with gas. A poppet valve at the accumulator base prevents the bladder extruding out into the piping.

Accumulators are sized by Boyle's law and a knowledge of the demands of the actuators. For the example system of Figure 6.2,

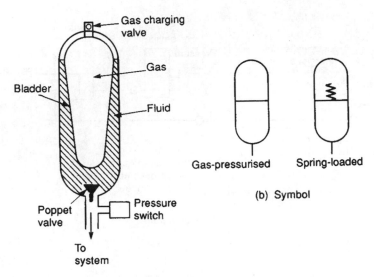

Figure 6.4   *The accumulator*

assuming a precharge of 120 bar, a charged accumulator pressure of 180 bar and a fall to a pressure to 160 bar during the removal of 17 litres of fluid: let V be volume of accumulator. This gives us the three states illustrated in Figure 6.5 to which Boyle's law can be applied to find the required accumulator volume.

From Figure 6.5b and c using Boyle's law:

$$160v = 180(v - 17)$$

which reduces to:

$$v = 153 \text{ litres}$$

From Figure 6.5a:

$$120V = 160 \times 153$$

or:

$$V = 204 \text{ litres}$$

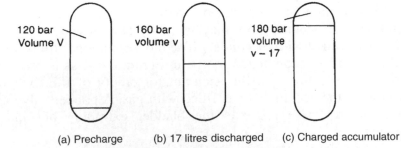

Figure 6.5   *Sizing an accumulator*

Hence an accumulator of around 250 litres is required, with a precharge of 120 bar and a pressure switch set at 180 bar.

Accumulators can also be used to act as 'buffers' on a system to absorb shocks and snub pressure spikes. Again the accumulator acts in similar manner to a capacitor in an electronic circuit.

An accumulator, however, brings an additional danger into the system, as it is possible for high pressures to exist in the circuit even though the pump has been stopped. If a coupling is opened under these circumstances the accumulator discharges all its fluid at working pressure. The author speaks from personal experience of having committed this cardinal sin and being covered in oil for his mistake!

Extreme care should therefore be taken when working on circuits with accumulators. Normally a manual or automatic blowdown valve is included to allow the accumulator pressure to be released. The pressure gauge should be observed during blowdown and no work undertaken until it is certain all pressure has been released. Figure 6.6 shows typical blowdown circuits.

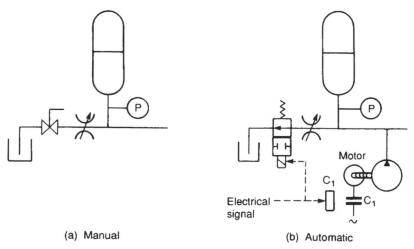

(a) Manual   (b) Automatic

Figure 6.6 *Accumulator blowdown circuits. In each case flow from the accumulator is restricted to prevent an explosive decompression*

Once a system has warmed up, a quick check can be made on the state of an accumulator with the flat of the hand. There should always be a significant temperature difference between the gas and the hydraulic oil and the oil/gas split can be detected by the temperature change on the body of the accumulator. If the whole body is the same temperature something has gone severely wrong with the gas bladder.

An accumulator is a pressurised vessel and as such requires certification if it contains more than 250 bar.litres. It will require a recorded expert visual inspection every five years and a full volumetric pressure test every ten years.

Hydraulic coolers and heat exchangers

Despite the occasional use of heaters mentioned earlier, the problem with oil temperature is usually keeping it *down* to the required 50°C. In small systems. the heat lost through reservoir walls is sufficient to keep the oil cool, but in larger systems additional cooling is needed. Table 6.1 shows typical heat losses from various sizes of reservoirs. It should be noted that the relationship between volume and heat loss (surface area) is non-linear, because surface area increases as the square of the linear dimensions, whereas volume increases as the cube.

Table 6.1   Heat loss for various tank volumes. These are only approximate as few tanks are pure cubes

Vol (l)	L (m)	Surface area (m^2)	Heat loss (kW)
250	0.63	1.98	0.5
500	0.8	3.2	1.0
1,000	1.0	5.0	1.5
2,000	1.25	7.8	2.5
10,000	2.15	23.1	15.0

Based on cube tank where
$L^3 = v$

Surface area = $5 \times L^2$ (to allow for air gap at top and poor heat transfer from base)

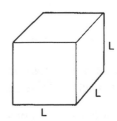

Heat loss approx 0.3 kW m^{-2}

Figure 6.7 shows two types of cooler and their symbols. Water cooling is most common and Figure 6.7a shows the usual form of a shell and tube heat exchanger which is fitted in the return line to the tank. Note that the cooling water flows in the opposite direction to the oil (giving rise to the term: counter-flow cooler). If the system is open to atmosphere and liable to stand unused in cold weather, protection must be included to prevent frost damage which can result in water-contaminated oil.

Air cooling is also common, shown in Figure 6.7b, with fans blowing air through a radiator matrix similar to those in motor cars (but, obviously, with a far higher pressure rating). Air cooling is noisy and occupies more space than a water cooler, but does not have the danger of contamination from leaks inside a water cooler.

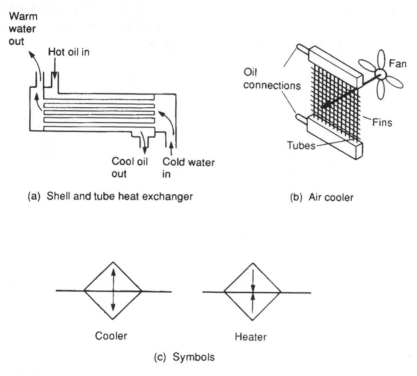

(a) Shell and tube heat exchanger   (b) Air cooler

Cooler   Heater

(c) Symbols

Figure 6.7 *Coolers and heat exchangers*

Hydraulic fluids

The liquid in a hydraulic system is used to convey energy and produce the required force at the actuators. Very early systems used water (in fact the name hydraulic implies water) but water has many disadvantages, the most obvious of which are its relatively high freezing point of 0°C, its expansion when freezing, its corrosive (rust formation) properties and poor lubrication. Modern fluids designed specifically for hydraulic circuits have therefore been developed.

The fluid conveys power in a hydraulic circuit, but it must also have other properties. Chapter 5 described the seals found in actuators. Moving parts in valves do not have seals; instead they rely on fine machining of spools and body to form the seal in conjunction with the fluid. Despite fine machining, irregularities still occur on the surface, shown in exaggerated form on Figure 6.8a. The fluid is required to pass between the two surfaces, holding them apart as Figure 6.8b, to reduce friction and prevent metal-to-metal contact which causes premature wear. Sealing and lubrication are therefore two important properties of hydraulic fluid.

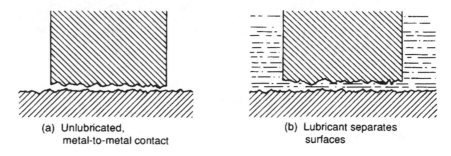

   (a) Unlubricated,   (b) Lubricant separates
      metal-to-metal contact   surfaces

Figure 6.8   *Need for lubrication from hydraulic fluid*

The temperature of hydraulic fluid tends to rise with the work done, an ideal operating temperature being around 50°C (a useful quick check is to touch pipes in a system: the hand can be left indefinitely on metal at 40°C; can touch metal at 50°C but long contact is distinctly uncomfortable; but cannot be left for more than a second or so on metal at 60°C. If you cannot touch the pipes, the oil is too hot!). The fluid must be able to convey heat from where it is generated (valves, actuators, frictional losses in pipes) and must not be affected itself by temperature changes.

The fluid can cause deterioration of components. An extreme case is water causing rust, but less obvious reactions occur. A water-glycol fluid, for example, attacks zinc, magnesium and cadmium – all fairly common materials. Some synthetic fluids interact with nitrile and neoprene, and special paint is needed on the inside of the reservoir with some fluids. The fluid must therefore be chosen to be compatible with the rest of the system.

The fluid itself comes under attack from oxygen in air. Oxidation of fluid (usually based on carbon and hydrogen molecules) leads to deleterious changes in characteristics and the formation of sludge or gum at low velocity points in the system. The resulting oxidation products are acidic in nature, leading to corrosion. The fluid of course must be chemically stable and not suffer from oxidation. The temperature of fluid strongly influences the rate of oxidation; which rises rapidly with increasing temperature.

The most common hydraulic fluid is petroleum based oil (similar to car engine oil) with additions to improve lubrication, reduce foaming and inhibit rust. With the correct additives it meets all the requirements and does not react adversely with any common materials.

Its one major disadvantage is flammability; petroleum oils readily ignite. Although few (if any) hydraulic systems operate at temperatures that could ignite the oil, a major leak *could* bring spilt oil into contact with an ignition source. The probability of leakage needs consideration if petroleum oils are to be used.

If safety dictates that a fire resistant fluid is required, an oil and water emulsion is commonly used (such fluids are also attractive on the grounds of cost). The most common form is a water-in-oil emulsion (roughly 40% water, 60% oil). Oil-in-water emulsions are sometimes used, but their lubricating properties are poor. Both types of mixture have a tendency to form rust and to foam, but these characteristics can be overcome by suitable additions. Both types also need regular checking to ensure the correct oil/water ratio is being maintained.

Another non-flammable fluid is a water/glycol mix. This consists of roughly equal proportions of water and glycol (similar to car antifreeze) plus additions to improve viscosity (see below), inhibit foaming and prevent rust to which water-based fluids are vulnerable. Glycol-based fluids interact with many common materials, so the system components must be carefully chosen.

High water content fluids (HWCF) use 95% water with 5% additives making them totally non-flammable. They are often called 95/5 micro emulsion. Their use needs some care as they have very low viscosity, for all practical purposes the same as water, making applications using them prone to leaks at joints and seals. Unlike normal fluids external leaks can be difficult to see as at the normal 40–50°C operating temperature the fluid evaporates away without leaving any trace.

Spool valves have an inherent leakage and this can be problematical with low viscosity fluids such as HWCF. Cartridge valves, described in Chapter 4, are therefore often used with HWCF.

The high water content makes precautions against rust very important. Any 95/5 components removed from service must be protected against exposure to air. Some manufacturers will not honour their warranties when 95/5 fluid has been used.

Synthetic fluids based on chemicals such as phosphate esters are also non-flammable and can be used at very high temperatures. These tend to have high densities, which limit the height allowed between tank and pump inlet without cavitation occurring, and do not operate well at low temperatures. Systems with synthetic fluids usually require heaters in the tank to preheat fluid to operating temperature. Synthetic fluids are the most expensive form of hydraulic oil.

The properties of a liquid are largely determined by its resistance to flow, which is termed its viscosity. In non-scientific terms we talk about treacle having high viscosity, and water having low viscosity. Both extremes bring problems; a low viscosity fluid flows easily and wastes little energy, but increases losses from leakage. A viscous fluid seals well, but is sluggish and leads to energy and pressure losses around the system. Hydraulic fluid has to hit a happy medium between these extremes, so some way of defining viscosity is required.

There are basically two techniques of specifying viscosity. The absolute scientific method measures the shear force between two plates separated by a thin fluid film, shown as Figure 6.9. The most common unit is the poise (a cgs unit) which is the measure of shear force in dynes, for surface areas of 1 cm² separated by 1 cm of fluid. The centipoise (0.01 poise) is a more practical unit. Kinematic viscosity, defined with a unit called the stokes, is given by the absolute viscosity (in poise) divided by the density (in gm cm⁻³).

A practical unit is the centi-stokes; a typical hydraulic fluid will have a viscosity of around 40 centi-stokes and low viscosity fluid such as HWCF about 1 centi-stoke. Not surprisingly this much lower viscosity means HWCF is very prone to leaks.

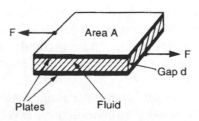

Figure 6.9   *Scientific definition of viscosity in terms of shear force*

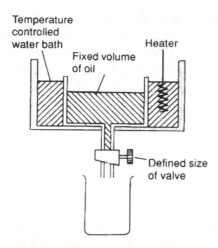

Figure 6.10 *Practical definition of viscosity*

The poise and the stokes are units denoting scientific definitions of viscosity. In hydraulics, all that is really needed is a relative comparison between different liquids. This is achieved with the practical experiment shown in Figure 6.10, where a fixed volume of oil is heated to a test temperature then allowed to drain out through a fixed-sized valve. The time taken to drain in seconds is a measure of the viscosity (being high for high viscosity liquids and low for low viscosity liquids).

The test of Figure 6.10 (generally performed at 100°F and 210°F with a volume of 60 cm³) gives viscosity in saybolt universal seconds (SUS). The Fahrenheit basis of these definitions come from the American origin. Hydraulic fluid normally has a viscosity between 150 and 250 SUS defined at 100°F, although higher values are used in high temperature applications.

Viscosity can also be given by similar tests for engine oils devised by the American Society of Automotive Engineers (SAE). These give Winter numbers with suffix W (e.g., 10W, 20W) defined at 0°F, and Summer numbers defined at 210°F. An oil rating of 10W SAE, for example, covers the range 6,000 to 12,000 SUS at 0°F, while 30SAE covers the range 58 to 70 SUS at 210°F.

Viscosity decreases with increasing temperature, and this is given in SAE units in the form SAE 10W50, for example. This variation in viscosity with temperature is defined by the viscosity index, a unit based on an arbitrary scale from zero (poor, large variation in viscosity with temperature) to 100 (good, small variation with temperature). The range zero to 100 was chosen to relate to standards obtainable with practical fluids rather than some absolute measurable standard. Most hydraulic oils have a viscosity index of about 90.

The reliability of a hydraulic system is strongly influenced by the state of fluid. Contamination from dirt or the products of oxidation and deterioration of a fluid's lubrication ability will lead to rapid wear and failure.

Pneumatic piping, hoses and connections

The various end devices in a pneumatic system are linked to the air receiver by pipes, tubes or hoses. In many schemes the air supply is installed as a fixed service similar, in principle, to an electrical ring main allowing future devices to be added as required. Generally, distribution is arranged as a manifold (as Figure 6.11a) or as a ring main (as Figure 6.11b). With strategically placed isolation valves, a ring main has the advantage that parts of the ring can be isolated for maintenance, modification or repair without affecting the rest of the system.

Pneumatic systems are vulnerable to moisture and, to provide drainage, the piping should be installed with a slope of about 1% (1 in 100) down from the reservoir. A water trap fitted at the lowest point of the system allows condensation to be run off, and all tap-offs are taken from the top of the pipe (Figure 6.11c) to prevent water collecting in branch lines.

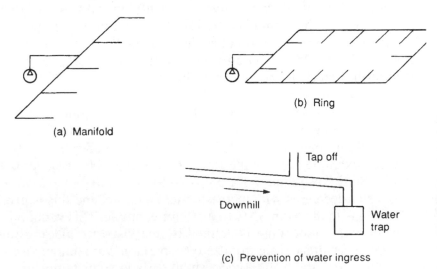

(a) Manifold

(b) Ring

(c) Prevention of water ingress

Figure 6.11   *Pneumatic piping*

The pipe sizing should be chosen to keep the pressure reasonably constant over the whole system. The pressure drop is dependent on maximum flow, working pressure, length of line, fittings in the line (e.g., elbows, T-pieces, valves) and the allowable pressure drop. The aim should be to keep air flow non-turbulent (laminar or streamline flow). Pipe suppliers provide tables or nomographs linking pressure drops to pipe length and different pipe diameters. Pipe fittings are generally specified in terms of an equivalent length of standard pipe (a 90 mm elbow, for example, is equivalent in terms of pressure drop to 1 metre of 90 mm pipe). If an intermittent large load causes local pressure drops, installation of an additional air receiver by the load can reduce its effect on the rest of the system. The local receiver is serving a similar role to a smoothing capacitor in an electronic power supply, or an accumulator in a hydraulic circuit.

If a pneumatic system is installed as a plant service (rather than for a specific well-defined purpose) pipe sizing should always be chosen conservatively to allow for future developments. Doubling a pipe diameter gives four times the cross-sectional area, and pressure drops lowered by a factor of at least ten. Retrofitting larger size piping is far more expensive than installing original piping with substantial allowance for growth.

Black steel piping is primarily used for main pipe runs, with elbow connections where bends are needed (piping, unlike tubing, cannot be bent). Tubing, manufactured to a better finish and more accurate inside and outside diameters from drawn or extruded flexible metals such as brass, copper or aluminium, is used for smaller diameter lines. As a very rough rule, tubing is used below 25 mm and piping above 50 mm – diameters in between are determined by the application. A main advantage of tubing is that swept angles and corners can be formed with bending machines to give simpler and leak-free installations, and minimising the pressure drops associated with fittings.

Connections can be made by welding, threaded connections, flanges or compression tube connectors. (Examples of compression fittings are illustrated in Figure 6.12.)

Welded connections are leak-free and robust, and are the prime choice for fixed main distribution pipe lines. Welding does, however, cause scale to be deposited inside the pipe which must be removed before use.

Threaded pipe connections must obviously have male threads on the pipes, and are available to a variety of standards, some of which

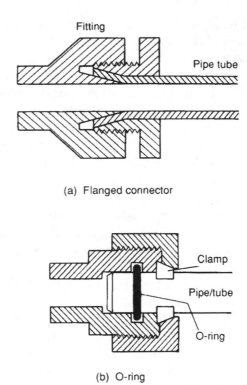

(a) Flanged connector

(b) O-ring

Figure 6.12   *Compression fittings*

are NPT (American National Pipe Threads), UNF (Unified
Pipe Threads), BSP (British Standard Pipe Threads) and Metric
Pipe Threads. The choice between these is determined by the
standards already chosen for a user's site. Taper threads are cone
shaped and form a seal between the male and female parts as they
tighten, with assistance from a jointing compound or plastic tapes.
Parallel threads are cheaper, but need an O-ring to provide the
seal.

A pipe run can be subject to shock loads from pressure changes
inside the pipe, and there can also be accidental outside impacts.
Piping must therefore be securely mounted and protected where
there is a danger from accidental damage. In-line fittings such as
valves, filters and treatment units should have their own mounting
and not rely on piping on either side for support.

At the relatively low pressure of pneumatic systems, (typically 5
to 10 bar), most common piping has a more than adequate safety
margin. Pipe strength should, however, be checked – as a burst air
line will scatter shrapnel-like fragments at high speed.

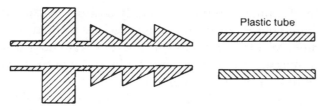

Plastic tube

Figure 6.13 *Barbed connector for plastic tube*

Plastic tubing is used for low pressure (around 6 bar) lines where flexibility is needed. Plastic connections are usually made with barbed push-on connectors, illustrated in Figure 6.13.

Where flexibility is needed at higher pressure, hosing can be used. Pneumatic hoses are constructed with three concentric layers; an inner tube made of synthetic rubber surrounded by a reinforcement material such as metal braiding. A plastic outer layer is then used to protect the hosing from abrasion.

Hose fittings need care in use, as they must clamp tightly onto the hose, but not so tightly as to cut through the reinforcement. Quick-disconnect couplings are used where hoses are to be attached and disconnected without the need of shut-off valves. These contain a spring-loaded poppet which closes the outlet when the hose is removed. There is always a brief blast of air as the connection is made or broken, which can eject any dirt around the connector at high speed. Extreme care must therefore be taken when using quick-disconnect couplings.

Hydraulic piping, hosing and connections

The differences between hydraulic and pneumatic piping primarily arise from the far higher operating pressures in a hydraulic system.

Particular care has to be taken to check the pressure rating of pipes, tubing, hosing and fittings, specified as the bursting pressure. A safety factor is defined as:

$$\text{safety factor} = \frac{\text{bursting pressure}}{\text{working pressure}}$$

Up to 60 bar, a safety factor of eight should be used, between 60 and 150 bar a safety factor of six is recommended, while above 150 bar a safety factor of four is required. This may be compared with pneumatic systems where safety factors of around 40 are normally obtained with simple standard components.

The choice of piping or tubing is usually a direct consequence of pressure rating. These can be manufactured as welded, or drawn (seamless) pipe. Welded pipe has an inherent weakness down the welded seam, making seamless pipes or tubing the preferred choice for all but the lowest pressure hydraulic systems.

Hydraulic piping is specified by wall thickness (which determines the pressure rating) and outside diameter (OD, which determines the size of fittings to be used). It follows that for a given OD, a higher pressure pipe has a smaller inside diameter (ID). American piping is manufactured to American National Standards Institute (ANSI) specifications, which define 10 sets of wall thickness as a schedule number from 10 to 160. The higher the number, the higher the pressure rating. 'Standard' piping is schedule 40.

Pipes should be sized to give a specified flow velocity according to the expected flow. Typical flow velocities are 7–8 m/sec for a pressure line, and 3–4 m/sec for a return line. The lower velocity is specified for the return line to reduce the back pressure. For a similar reason the velocity in a pump suction line should be in the range 1.5–2 m/sec. At the point of exit from the return line diffuser into the tank the velocity should be very low, below 0.3 m/sec, to prevent stirring up any contamination at the base of the tank.

Like pneumatic piping, joints can be made by welding, with compression fittings (similar to those in Figure 6.12 but of higher pressure rating) or threaded connections and flanges. Particular care needs to be taken to avoid leaks at joints; in pneumatic systems a leak leads to loss of downstream pressure and perhaps an objectionable noise whereas a hydraulic leak loses expensive fluid and creates an oil-pool which is a fire and safety hazard.

Flexible hosing is constructed in several concentric layers, with the inner tubing being chosen to be compatible with the hydraulic fluid and its temperature. One (or more) braided reinforcing layers are used. At higher pressures the braiding will be wire. The outer layer is designed to resist abrasion and protect the inner layers. Hoses are generally manufactured complete with fittings. Hydraulic hoses, like pneumatic hoses, must be installed without twists (which can lead to failure at the fittings).

Quick-disconnect hydraulic connections are available, but the higher pressure, risk of spillage and danger of introducing dust into the system restricts their usage.

7

Process control pneumatics

If some industrial process is to be automatically controlled, there will be many process variables (e.g. temperature, flow, pressure, level) which need to be measured and kept at the correct value for safety and economical operation. In Figure 7.1, for example, water flow in a pipe is to be kept at some preset value.

In Figure 7.1 the flow is measured to give the current value (usually termed PV – for process variable). This is compared with the required flow (called SP – for set point) to give an error signal, which is passed to a controller. This adjusts the actuator drive signal to move the valve in the direction to give the required flow (i.e. PV = SP, giving zero error). The arrangement of Figure 7.1 is called *closed loop control* because a loop is formed by the controller, actuator and measuring device.

In many plants, closed loop control is achieved by electronics, or even computer, techniques with the various signals represented by electric currents. A common standard uses a current within the

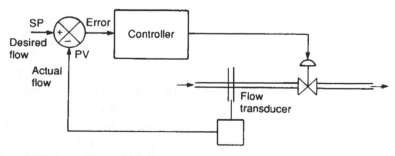

Figure 7.1   *Closed loop control*

range 4 to 20 mA. If this represents a water flow from 0 to 1500 1 min^{-1}, for example, a flow of 1000 1 min^{-1} is represented by a current of 14.67 mA.

Electrical representation, and electronic devices, are not the only possibility, however. Process control history goes back before the advent of electronics (some early examples being speed governors on steam engines and an early servosystem for ships' rudders designed by Isambard Kingdom Brunel). Much of the original process control work was based around pneumatic devices, with the various signals represented by pneumatic pressures.

Perhaps surprisingly, pneumatic process control has by no means been superseded by electronic and microprocessor technology, so it is worth looking at the reasons for its popularity. First and foremost is safety. Much process control is done in chemical or petrochemical plants where explosive atmospheres are common. If electrical signals are used, great care must be taken to ensure no possible fault can cause a spark, which could ignite an explosive atmosphere. While this can be achieved, the result is complex and maintenance may be difficult (test instruments must also be classified safe for use in an explosive atmosphere).

A pneumatic system contains only air, so it presents no hazard under these conditions. No particular care needs to be taken with installation, and maintenance work can be carried out 'live' with simple non-electrical test instruments.

A great deal of design and application experience has evolved over the years, and this base of knowledge is another major reason for the continuing popularity of pneumatic control. Companies with a significant investment in pneumatic control and a high level of staff competency are unlikely to change.

Many devices in the loop are, in any case, best provided by pneumatic techniques. Although electrical actuators are available, most valves are driven by pneumatic signals – even when transducer and controller are electronic.

Signals and standards

Signals in process control are generally represented by a pressure which varies over the range 0.2 to 1.0 bar or the almost identical imperial equivalent 3 to 15 psig. If the water flow of 0 to 1500 1 min^{-1} is represented pneumatically, 01 min^{-1} is shown by a pressure of 0.2 bar, 1500 1 min^{-1} is 1.0 bar, while 10001 min^{-1} is 0.733 bar.

The lower range pressure of 0.2 bar (3 psig in the imperial range) is known as an offset zero and serves two purposes. First is to warn about damage to signal lines linking the transmitter and the controller or indicator (the 4 mA offset zero of electrical systems also gives this protection). In Figure 7.2a a pneumatic flow transmitter is connected to a flow indicator. A pneumatic supply (typically, 2 to 4 bar) is connected to the transmitter to allow the line pressure to be raised. The transmitter can also vent the line to reduce pressure (corresponding to reducing flow). If the line is damaged it is probably open to atmosphere giving a pressure of 0 bar, regardless of the transmitter's actions. As the indicator is scaled for 0.2 to 1 bar, a line fault therefore causes the indicator to go offscale, negatively. Loss of the pressure supply line causes a similar fault indication.

The offset zero also increases the speed of response. In Figure 7.2b a sudden increase in flow is applied to the transmitter at time A. The flow transmitter connects the supply to the line, causing an

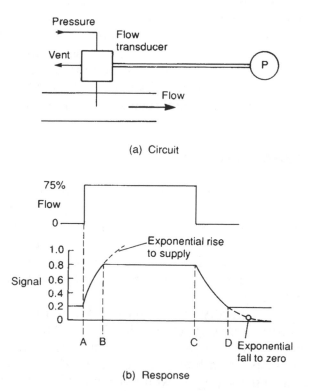

(a) Circuit

(b) Response

Figure 7.2 *Advantage of an offset zero*

exponential increase in pressure (with a time constant determined by the line volume). The pressure rises towards the supply pressure, but at time B the correct pressure of 0.8 bar is reached, and the transmitter disconnects the supply.

The pressure stays at 0.8 bar until time C, when the flow rapidly falls to zero. The transmitter vents the line and the pressure falls exponentially towards 0 bar (with time constant again determined by line volume). At time D, a pressure of 0.2 bar is reached (corresponding to zero flow) and the transmitter stops venting the line. For increasing indication, the offset zero has little effect, but for decreasing indication, the transmitter would need to completely vent the line without an offset zero to give zero indication. With a first order lag response, this will theoretically take an infinite time, but even with some practical acceptance of error, time CD will be significantly extended.

Speed of response is, in any case, the Achilles heel of pneumatic signals. With an infinitely small time constant (given by zero volume lines), the best possible response can only be the speed of sound (330m s^{-1}). If signal lines are over a hundred metres or so in length, this transit delay is significant. To this is added the first order lag caused by the finite volume of the line, and the finite rate at which air flows into or out of the line under transmitter control. For a fast response, line volume must be small (difficult to achieve with long lines) and the transmitter able to deliver, or vent large flow rates. In practice, time constants of several seconds are quite common.

The flapper-nozzle

Most properties (eg flow, pressure, level, error, desired valve position) can be converted to a small movement. The heart of all pneumatic process control devices is a device to convert a small displacement into a pressure change, which represents the property causing the displacement. This is invariably based on the flapper-nozzle, whose arrangement, characteristic and application are illustrated in Figure 7.3.

An air supply (typically, 2 to 4 bar) is applied to a very fine nozzle via a restriction as shown in Figure 7.3a. The signal output side of the nozzle feeds to a closed (non-venting) load, such as an indicator. Air escapes as a fine jet from the nozzle, so the pressure at A is lower than the supply pressure because of the pressure drop across the restriction.

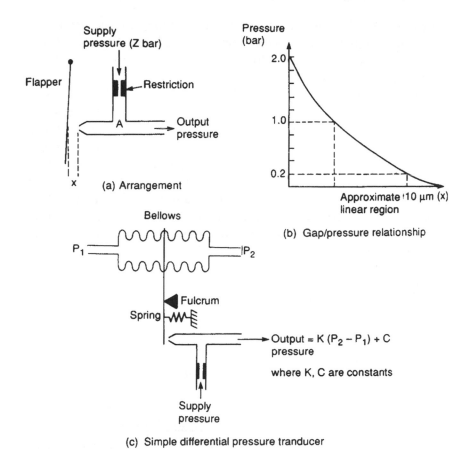

(a) Arrangement

(b) Gap/pressure relationship

(c) Simple differential pressure tranducer

Figure 7.3 *The flapper-nozzle, the basis of pneumatic process control*

Air loss from the jet (and hence pressure at A) is influenced by the gap x between the nozzle and movable flapper; the smaller the gap, the lower the air flow and higher the pressure. A typical response is shown in Figure 7.3b, illustrating the very small range of displacement and the overall non-linear response. The response can, however, be considered linear over a limited range (as shown) and the flapper-nozzle is generally linearised by use of a force balance system as described later.

Figure 7.3c shows a very simple differential pressure transducer which may be used as a flow transmitter by measuring the pressure drops across an orifice plate. The difference in pressure between P_1 and P_2 causes a force on the flapper. Assuming $P_1 > P_2$ (which is true for the direction of flow shown), the top of the flapper is pushed to the right until the force from (P_1–P_2) is matched by the force from the spring extension. Flapper nozzle gap, and hence the output pressure, is thus determined by the differential pressure and the flow through the orifice plate.

The arrangement of Figure 7.3c is non-linear, and incapable of maintaining output pressure to a load with even a small loss of air. Even with a totally sealed load the minimal air flow through the restriction leads to a first order lag response with a very long time constant. A flapper-nozzle is therefore usually combined with an air amplifier, or volume booster, which takes a pressure as the input and gives a linearly-related pressure output – with an ability to supply a large volume of air. When combined with the force balance principle described later, the inherent non-linearity of the flapper nozzle can be overcome.

Volume boosters

An air amplifier is illustrated in Figure 7.4. It is provided with an air supply (typically 2-4 bar) and an input signal pressure. The amplifier admits air to, or vents air from, the output to maintain a

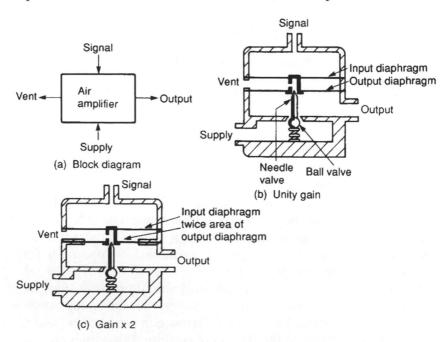

Figure 7.4   *Volume boosters or air amplifiers*

constant output/input ratio. An amplifier with a gain of two, for example, turns a 0.2 to 1 bar signal range to a 0.4 to 2 bar range. Output pressure, controlled by the amplifier, has the ability to provide a large air volume and can drive large capacity loads.

A unity gain air amplifier is shown in Figure 7.4b. It consists of two equal-area linked diaphragms, which together operate a needle and ball valve arrangement. The low volume input signal is applied to the upper diaphragm and the output pressure to the lower diaphragm. If output pressure is lower than inlet pressure, the diaphragm is pushed down, closing needle valve and opening ball valve to pass supply air to the load and increase output pressure.

If the output pressure is high, the diaphragm is forced up, closing the spring-loaded ball valve and opening the needle valve to allow air to escape through the vent and reduce output pressure. The amplifier stabilises with output and input pressures equal.

The input port has a small and practically constant volume, which can be controlled directly by a flapper-nozzle. The output pressure tracks changes in inlet pressure, but with the ability to supply a large volume of air.

An air amplifier balances when forces on the two diaphragms are equal and opposite. Equal area diaphragms have been used in the unity gain amplifier of Figure 7.4b. The area of the input diaphragm in the amplifier of Figure 7.4c is *twice* the area of the output diaphragm. For balance, the output pressure must be twice the input pressure, giving a gain of two. In general, the amplifier gain is given by:

$$\text{gain} = \frac{\text{input area}}{\text{output area}}$$

The air relay and the force balance principle

Air amplifiers balance input pressure and output pressure. An air relay, on the other hand (illustrated in Figure 7.5), balances input pressure with the force from a range spring. An increasing input signal causes air to pass from the supply to the load, while a decreasing input signal causes air to vent from the load. In the centre of the input signal range, there is no net flow to or from the output port.

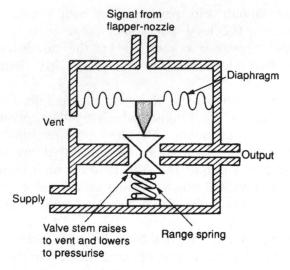

Figure 7.5   *The air relay*

An air relay is used to linearise a flapper-nozzle, as shown in Figure 7.6. Here, force from the unbalance in input pressures P_1 and P_2 is matched exactly by the force from the feedback bellows whose pressure is regulated by the air relay.

Suppose flow in the pipe increases, causing pressure difference P_1-P_2 to increase. Increased force from the bellows at the top decreases the flapper gap causing pressure at the air relay input to

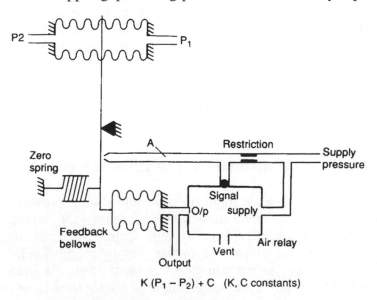

$$K\,(P_1 - P_2) + C \quad \text{(K, C constants)}$$

Figure 7.6   *The force balance principle*

rise. This causes air to pass to the feedback bellows, which apply a force opposite to that from the signal bellows.

The system balances when the input pressure from the flapper nozzle to the air relay (point A) is at the centre of its range at which point the air relay neither passes air nor vents the feedback bellows. This corresponds to a fixed flapper-nozzle gap.

Figure 7.6 thus illustrates an example of a feedback system where the pressure in the feedback bellows is adjusted by the air relay to maintain a constant flapper-nozzle gap. The force from the feedback bellows thus matches the force from the input signal bellows, and output pressure is directly proportional to (P_1-P_2). The output pressure, driven directly from the air relay, can deliver a large air volume.

The arrangement in Figure 7.6 effectively operates with a fixed flapper-nozzle gap. This overcomes the inherent non-linearity of the flapper-nozzle. It is known as the force balance principle and is the basis of most pneumatic process control devices.

Pneumatic controllers

Closed loop control, discussed briefly earlier, requires a controller which takes a desired (set point) signal and an actual (process variable) signal, computes the error then adjusts the output to an actuator to make the actual value equal the desired value.

The simplest pneumatic controller is called a proportional only controller, shown schematically in Figure 7.7. The output signal here is simply the error signal multiplied by a gain:

$$OP = K \times error$$
$$= K \times (SP - PV). \quad (7.1)$$

where K is the gain.

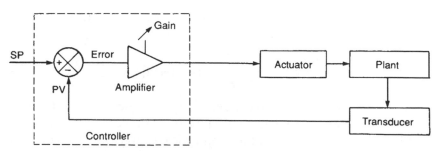

Figure 7.7 *Proportional only controller*

Comparison of the controller in Figure 7.7 with the force balance transmitter in Figure 7.6 shows that the differential pressure measurement (P_1-P_2) performs the same function as error subtraction (SP–PV). We can thus construct a simple proportional only controller with the pneumatic circuit of Figure 7.6. Gain can be set by moving the pivot position.

The output of a proportional controller is simply $K \times$ error, so to get any output signal, an error signal must exist. This error, called the offset, is usually small, and can be decreased by using a large gain. In many applications, however, too large a gain causes the system to become unstable.

In these circumstances a modification to the basic controller is used. A time integral of the error is added to give:

$$OP = K \left(\text{error} + \frac{1}{T_i} \int \text{error dt} \right). \qquad (7.2)$$

Controllers following expression 7.2 are called proportional plus integral (P+I) controllers, illustrated in Figure 7.8. The constant T_i, called the integral time, is set by the user. Often the setting is given in terms of $1/T_i$ (when the description repeats/min is used). A controller following expression 7.2 has a block diagram shown in Figure 7.8a, and responds to a step response as shown in Figure 7.8b. As long as an error exists, the controller output creeps up or down to a rate determined by T_i. Only when there is no error is the controller output constant. Inclusion of the integral term in expression 7.2 removes the offset error.

A pneumatic P+I controller can be constructed as shown in Figure 7.8c. Integral bellows oppose the action of the feedback bellows, with the rate of change of pressure limited by the T_i setting valve. The controller balances the correct flapper-nozzle gap to give zero error, with PV=SP and equal forces from the integral and feedback bellows.

A further controller variation, called the three term or P+I+D, controller uses the equation

$$OP = K \left(\text{error} + \frac{1}{T_i} \int \text{error dt} + T_d \frac{\text{d error}}{\text{dt}} \right). \qquad (7.3)$$

where T_d is a user-adjustable control, called the derivative time. Addition of a derivative term makes the control output change quickly when SP or PV are changing quickly, and can also serve to make a system more stable.

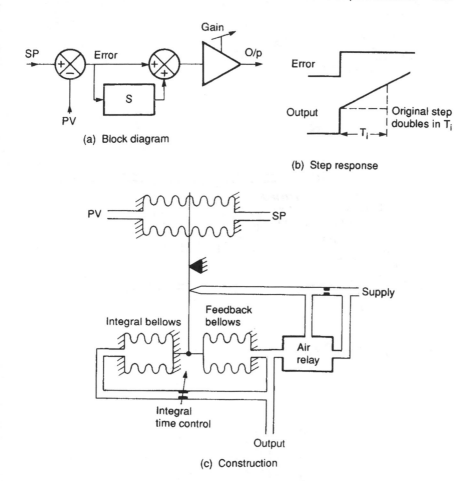

(a) Block diagram

(b) Step response

(c) Construction

Figure 7.8 *Proportional plus integral (P + I) controller*

Pneumatic three term control can be achieved with the arrangement of Figure 7.9, where the action of the feedback bellows has been delayed. The three user adjustable terms in expression 7.3 (gain K, integral time T_i, derivative time T_d) are set by beam pivot point and two bleed valves to give the best plant response. These controls do, however, interact to some extent – a failing not shared by electronic controllers.

Figure 7.10 represents the typical front panel of a controller. Values of SP, PV and controller output are displayed and the operator can select between automatic and manual operation. The desired value (SP) can be adjusted in auto or the controller output set directly in manual. The operator does not have access to K, T_i, T_d setting controls; these are adjusted by the maintenance technician.

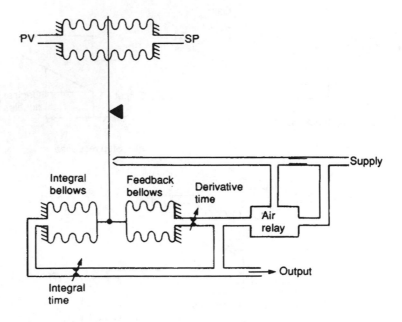

Figure 7.9 *Three term (P + I + D) controller*

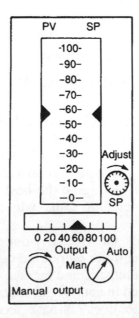

Figure 7.10 *Front panel of a typical controller*

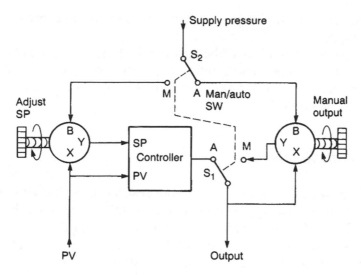

Figure 7.11 *Internal arrangement giving bumpless transfer*

Internally the controller is arranged as shown in Figure 7.11. Setpoint and manual output controls are pressure regulators, and the auto/manual switch simply selects between the controller and manual output pressures. If the selection, however, just switched between P_c and P_m there would be a step in the controller output. The pressure regulators are designed so their output Y tracks input X, rather than the manual setting when a pressure signal is applied to B. The linked switch S_2 thus makes the setpoint track the process variable in manual mode, while manual output P_m tracks the controller output in automatic mode. 'Bumpless' transfers between automatic and manual can therefore be achieved.

Process control valves and actuators

In most pneumatic process control schemes, the final actuator controls the flow of a fluid. Typical examples are liquid flow for chemical composition control, level control, fuel flow for temperature control and pressure control. In most cases the actual control device will be a pneumatically actuated flow control valve.

Even with totally electronic or computer-based process control schemes, most valves are pneumatically-operated. Although electrically-operated actuators *are* available, pneumatic devices tend to be cheaper, easier to maintain and have an inherent, and predictable, failure mode.

It is first useful to discuss the way in which fluid flow can be controlled. It is, perhaps, worth noting that these devices give full proportional control of fluid flow, and are *not* used to give a simple flow/no-flow control.

Flow control valves

All valves work by putting a variable restriction in the flow path. There are three basic types of flow control valves, shown in Figures 7.12 to 7.14. Of these the plug, or globe valve (Figure 7.12) is probably most common. This controls flow by varying the vertical plug position, which alters the size of the orifice between the tapered plug and valve seat. Normally the plug is guided and constrained from sideways movement by a cage, not shown in Fig. 7.12a for simplicity.

The valve characteristics define how the valve opening controls flow. The characteristics of the globe valve can be accurately predetermined by machining the taper of the plug. There are three common characteristics, shown in Figure 7.12b. These are specified for a constant pressure drop across the valve, a condition which rarely occurs in practical plants. In a given installation, the flow through a valve for a given opening depends not only on the valve, but also on pressure drops from all the other items and the piping in the rest of the system. The valve characteristic (quick opening, linear, or equal percentage) is therefore chosen to give an approxi-

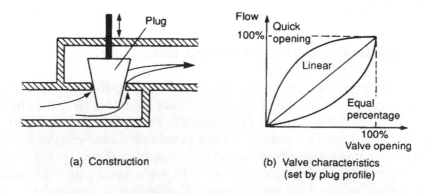

(a) Construction

(b) Valve characteristics
(set by plug profile)

Figure 7.12   *The plug valve*

mately linear flow/valve position relationship for this particular configuration.

A butterfly valve, shown in Figure 7.13, consists of a large disc which is rotated inside the pipe, the angle determining the restriction. Butterfly valves can be made to any size and are widely used for control of gas flow. They do, however, suffer from rather high leakage in the shut-off position and suffer badly from dynamic torque effects, a topic discussed later.

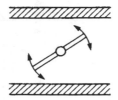

Figure 7.13   *The butterfly valve*

The ball valve, shown in Figure 7.14, uses a ball with a through hole which is rotated inside a machined seat. Ball valves have an excellent shut-off characteristic with leakage almost as good as an on/off isolation valve.

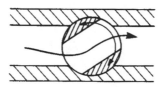

Figure 7.14   *The ball valve*

When fluid flows through a valve, dynamic forces act on the actuator shaft. In Figure 7.15a, the flow assists opening (and opposes the closing) of the valve. In Figure 7.5b, the flow assists the closing (and opposes the opening) of the valve. The latter case is particularly difficult to control at low flows as the plug tends to slam into the seat. This effect is easily observed by using the plug and chain to control flow of water out of a household bath.

The balanced valve of Figure 7.15c uses two plugs and two seats with opposite flows and gives little dynamic reaction onto the actuator shaft. This is achieved at the expense of higher leakage, as manufacturing tolerances cause one plug to seat before the other.

Butterfly valves suffer particularly from dynamic forces, a

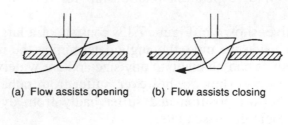

(a) Flow assists opening   (b) Flow assists closing

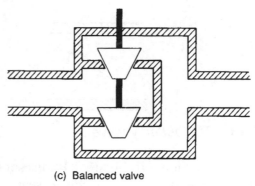

(c) Balanced valve

Figure 7.15   *Dynamic forces acting on a valve*

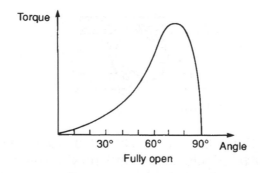

Figure 7.16   *Torque on a butterfly valve*

typical example being shown in Figure 7.16. As can be seen, maximum force occurs just before the fully open position, and this force acts to open the valve. It is not unknown for an actuator to be unable to move a butterfly valve off the fully open position and it is consequently good practice to mechanically limit opening to about 60°.

Actuators

The globe valve of Figure 7.12 needs a linear motion of the valve stem to control flow, whereas the butterfly valve of Figure 7.13 and the ball valve of Figure 7.14 require a rotary motion. In practice all, however, use a linear displacement actuator – with a mechanism similar to that in Figure 7.17 used to convert a linear stroke to an angular rotation if required.

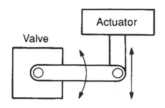

Figure 7.17 *Conversion from linear actuator motion to rotary valve motion*

Pneumatic valve actuators are superficially similar to the linear actuators of Chapter 5, but there are important differences. Linear actuators operate at a constant pressure, produce a *force* proportional to applied pressure and are generally fully extended or fully retracted. Valve actuators operate with an applied pressure which can vary from, say, 0.2 to 1 bar, producing a *displacement* of the shaft in direct proportion to the applied pressure.

A typical actuator is shown in Figure 7.18. The control signal is applied to the top of a piston sealed by a flexible diaphragm. The downward force from this pressure ($P \times A$) is opposed by the spring compression force and the piston settles where the two forces are equal, with a displacement proportional to applied pressure. Actuator gain (displacement/pressure) is determined by the stiffness of the spring, and the pressure at which the actuator starts to move (0.2 bar say) is set by a pre-tension adjustment.

Figure 7.18b illustrates the action of the rubber diaphragm. This 'peels' up and down the cylinder wall so the piston area remains constant over the full range of travel.

The shaft of the actuator extends for increasing pressure, and fails in a fully up position in the event of the usual failures of loss of air supply, loss of signal or rupture of the diaphragm seal. For this reason such an actuator is known as a fail-up type.

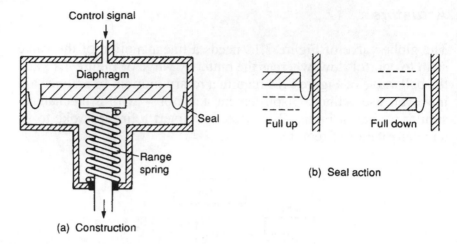

(a) Construction

(b) Seal action

Figure 7.18 *Fail-up actuator*

In the actuator of Figure 7.19, on the other hand, signal pressure is applied to the bottom of the piston and the spring action is reversed. With this design the shaft moves up for increasing pressure and moves down for common failure modes. This is known as a fail-down or reverse acting actuator.

One disadvantage of this design is the need for a seal on the valve shaft.

Where safety is important, valve and actuator should be chosen to give the correct failure mode. A fuel valve, for example, should fail closed, while a cooling water valve should fail open.

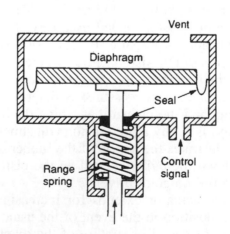

Figure 7.19 *Fail-down actuator*

Valve actuators tend to have large surface areas to give the required force, which means a significant volume of air is above the piston. Valve movement leads to changes in this volume, requiring air to be supplied from, or vented by, the device providing the pressure signal. A mismatch between the air requirements of the actuator and the capabilities of the device supplying pressure signal results in a slow, first order lag response.

The net force acting on the piston in Figures 7.18 and 7.19 is the sum of force from the applied pressure, the opposing spring force *and* any dynamic forces induced into the valve stem from the fluid being controlled. These dynamic forces therefore produce an offset error in valve position. The effect can be reduced by increasing the piston area or the operating pressure range, but there are limits on actuator size and the strength of the diaphragm seal. In Figure 7.20 a double-acting piston actuator operating at high pressure is shown. There is no restoring spring, so the shaft is moved by application of air to, or venting of air from, the two sides of the piston. A closed loop position control scheme is used, in which shaft displacement is compared with desired displacement (ie, signal pressure) and the piston pressures adjusted accordingly. The arrangement of Figure 7.20 is called a valve positioner, and correctly positions the shaft despite dynamic forces from the valve itself.

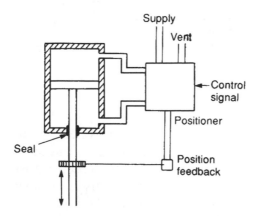

Figure 7.20   *Double-acting cylinder (holds position on failure)*

Valve positioners

A valve positioner is used to improve the performance of a pneumatically-operated actuator, by adding a position control loop around the actuator as shown in Figure 7.21. They are mainly used:

- to improve the operating speed of a valve;

- to provide volume boosting where the device providing the control signal can only provide a limited volume of air. As noted previously a mismatch between the capabilities of driver and the requirements of an actuator results in a first order lag response with a long time constant;

- to remove offsets resulting from dynamic forces in the valve (described in the previous section);

- where a pressure boost is needed to give the necessary actuator force;

- where a double-acting actuator is needed (which cannot be controlled with a single pressure line).

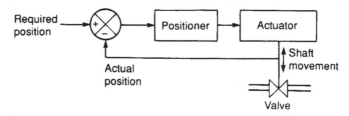

Figure 7.21   *The valve positioner*

There are two basic types of valve positioner. Figure 7.22 shows the construction of a valve positioner using a variation of the force balance principle described earlier. The actuator position is converted to a force by the range spring. This is compared with the force from the signal pressure acting on the input diaphragm. Any mismatch between the two forces results in movement of the beam and a change in the flapper-nozzle gap.

If the actuator position is low, the flapper-nozzle gap decreases, causing a rise in pressure at point A. This causes the spool to rise, connecting supply air to output 1, and venting output 2, resulting in the lifting of the actuator. If actuator position is high,

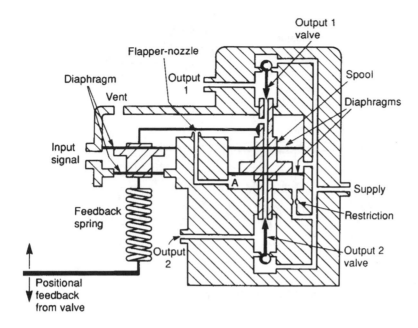

Figure 7.22 *Force balance valve positioner*

the flapper-nozzle gap increases and pressure at A falls causing the spool to move down applying air to output 2 and venting output 1, which results in the actuator lowering. The actuator thus balances when the range spring force (corresponding to actuator position) matches the force from the input signal pressure (corresponding to the required position) giving a constant flapper-nozzle gap.

The zero of the positioner is set by the linkage of the positioner to the valve shaft and the range by the spring stiffness. Fine zero adjustment can be made by a screw at the end of the spring.

The second type of positioner, illustrated in Figure 7.23, uses a motion balance principle. The valve shaft position is converted to a small displacement and applied to one end of the beam controlling the flapper-nozzle gap. The input signal is converted to a displacement at the other end of the beam. The pressure at A resulting from the flapper-nozzle gap is volume boosted by an air relay which passes air to, or vents air from, the actuator, to move the shaft until the flapper-nozzle gap is correct. At this point, the actuator position matches the desired position.

Positioners are generally supplied equipped with gauges to indicate supply pressure, signal pressure and output pressures, as illus-

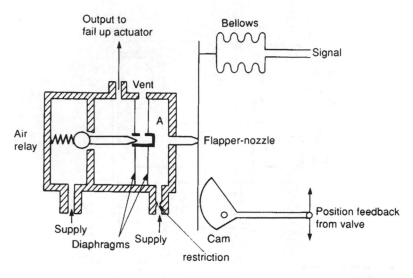

Figure 7.23 *Motion balance positioner*

trated in Figure 7.24 for a double-acting actuator. Often, bypass valves are fitted to allow the positioner to be bypassed temporarily in the event of failure with the signal pressure sent directly to the actuator .

Converters

The most common process control arrangement is probably electronic controllers with pneumatic actuators and transducers. Devices are therefore needed to convert between electrical analog

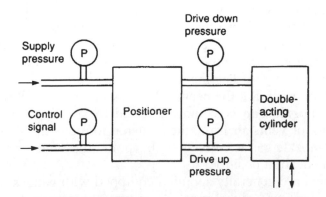

Figure 7.24 *Pressure indication on a positioner for fault finding*

signals and the various pneumatic standards. Electrical to pneumatic conversion is performed by an I–P converter, while pneumatic to electrical conversion is performed by a device called, not surprisingly, a P–I converter.

I–P converters

Figure 7.25 illustrates a common form of I–P converter based on the familiar force balance principle and the flapper-nozzle. Electrical current is passed through the coil and results in a rotational displacement of the beam. The resulting pressure change at the flapper-nozzle gap is volume-boosted by the air relay and applied as a balancing force by bellows at the other end of the beam. A balance results when the force from the bellows (proportional to output pressure) equals the force from the coil (proportional to input electrical signal).

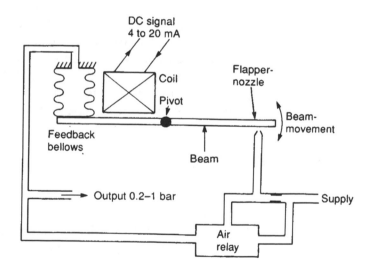

Figure 7.25 *Current to pressure (I–P) converter*

P–I converters

The operation of a P–I converter. illustrated in Figure 7.26 again uses the force balance principle. The input pressure signal is applied to bellows and produces a deflection of the beam. This deflection is measured by a position transducer such as an LVDT (linear variable

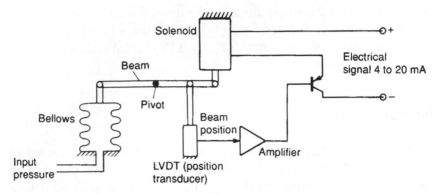

Figure 7.26   *Pressure to current (P–I) converter*

differential transformer). The electrical signal corresponding to the
deflection is amplified and applied as current through a coil to
produce a torque which brings the beam back to the null position.
At balance, the coil force (proportional to output current) matches
the force from the bellows (proportional to input signal pressure).

The zero offset (4 mA) in the electrical signal is sufficient to drive
the amplifier in Figure 7.26, allowing the two signal wires to also act
as the supply lines. This is known as two-wire operation. Most P–I
converters operate over a wide voltage range (eg, 15 to 30 V). Often,
the current signal of 4 to 20 mA is converted to a voltage signal
(commonly in the range 1 to 5 V) with a simple series resistor.

Sequencing applications

Process control pneumatics is also concerned with sequencing i.e.
performing simple actions which follow each other in a simple
order or with an order determined by sensors. Electrical equivalent
circuits are formed with relays, solid state logic or programmable
controllers.

A simple example of a pneumatic sequencing system is illustrated in Figure 7.27, where a piston oscillates continuously between
two striker-operated limit switches LS_1 and LS_2. These shift the
main valve V_1 with pilot pressure lines. The main valve spool has
no spring return and remains in position until the opposite signal is
applied. Shuttle valves V_2 and V_3 allow external signals to be
applied via ports Y and Z.

Time is often used to control a sequence (eg, feed a component,
wait five seconds, feed next component). A time delay valve is con-

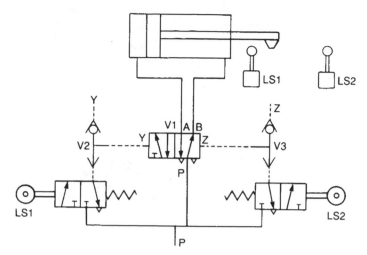

Figure 7.27 *A sequencing example; the cylinder oscillates between LS1 and LS2*

structed as illustrated in Figure 7.28a. Input signal X is a pilot signal moving the spool in main valve V_1, but it is delayed by the restriction valve and the small reservoir volume V. When X is applied, pilot pressure Y rises exponentially giving a delay T before the pilot operating pressure is reached. When X is removed, the non-return valve quickly vents the reservoir giving a negligible off delay. Figure 7.28b shows the response. As shown, the valve is a delay-on valve. If the non-return valve is reversed delay-off action is achieved.

Sequencing valves are used to tie pressure-controlled operations together. These act somewhat like a pilot-operated valve, but the

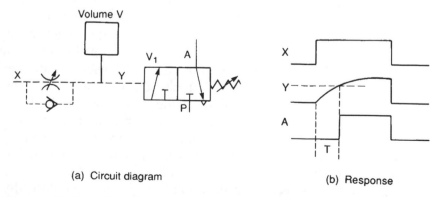

(a) Circuit diagram   (b) Response

Figure 7.28 *The time delay (see also Figure 4.28 for construc-*

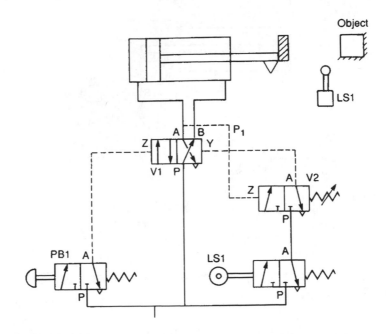

Figure 7.29   *Sequencing valve application*

designer can control the pressure at which the valve operates. A typical application is shown in Figure 7.29 where a cylinder is required to give a certain force to an object. Valve V_2 is the sequence valve and operates at a pressure set by the spring. The sequence is started by pushbutton PB_1, which shifts the pilot spool in the main valve V_1 causing the cylinder to extend. When the cylinder reaches full extension, limit switch LS_1 operates and pressure P_1 starts to rise. When the preset pressure is reached sequence valve V_2 operates, moving the spool in main valve V_1 and retracting the cylinder.

The two applications given so far have used limit switch operated valves to control sequences. Pneumatic proximity sensors can also be used. The reflex sensor of Figure 7.30 uses an annular nozzle jet of air the action of which removes air from the centre bore to give a light vacuum at the signal output X. If an object is placed in front of the sensor, flow is restricted and a significant pressure rise is seen at X. Another example is the interruptible jet sensor (Figure 7.31) which is simple in operation but uses more air. A typical application could be sensing the presence of a drill bit to indicate 'drill complete' in a pneumatically controlled machine tool. With no object present, the jet produces a

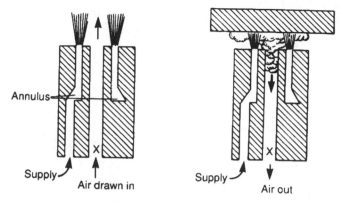

Figure 7.30 *Reflex proximity switch*

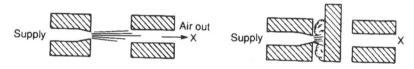

Figure 7.31 *Interruptible jet limit switch*

pressure rise at signal output X. An object blocking this flow, causes X to fall to atmospheric pressure.

With both types of sensor, air consumption can be a problem. To reduce air usage, low pressure and low flow rates are used. Both of these results in a low pressure signal at X which requires pressure amplification or low pressure pilot valves before it can be used to control full pressure lines.

Logic devices (AND, OR gates and memories) are part of the electrical tool kit for sequencing applications. The pneumatic equivalent (Figure 7.32) uses the wall attachment or Coanda effect. A fluid stream exiting from a jet with a Reynolds number in excess of 1500 (giving very turbulent flow) tends to attach itself to a wall and remain there until disturbed (Figure 7.32a).

This principle is used to give a pneumatic set/reset (S-R) flip-flop memory in Figure 7.32b. If the set input is pulsed, the flow attaches itself to the right-hand wall, exiting via output Q. If the set input is then removed the Coanda effect keeps the flow on this route until the reset input is pulsed.

Figure 7.32c shows a fluidic OR/NOR gate. A small bias pressure keeps the signal on the right-hand wall, which causes it to exit via

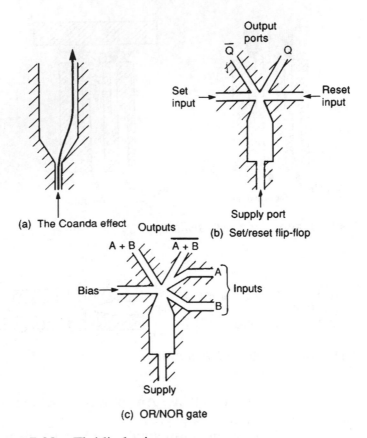

(a) The Coanda effect

(b) Set/reset flip-flop

(c) OR/NOR gate

Figure 7.32   Fluidic logic

the right-hand port. If signal A or B is applied (at higher pressure than the bias) the flow switches over to the (A+B) output. When both A and B signals are removed, the bias pressure switches the flow back again.

Logic functions can also be performed by series connections of valves (to give the AND operation) shuttle valves (to give the OR operation) and pilot-operated spools (to give flip-flop memories). Valve V_1 in Figure 7.27, for example, acts as an S-R flip-flop memory.

8

Safety, fault-finding and maintenance

Safety

Most industrial plant has the capacity to maim or kill. It is therefore the responsibility of all people, both employers and employees, to ensure that no harm comes to any person as a result of activities on an industrial site.

Not surprisingly, this moral duty is also backed up by legislation. It is interesting that most safety legislation is re-active, i.e. responding to incidents which have occurred and trying to prevent them happening again. A prime example of this is the CDM regulations which arose because of the appalling safety record in the construction industry.

Safety legislation differs from country to country, although harmonization is underway in Europe. This section describes safety from a British viewpoint, although the general principles apply throughout the European community and are applicable in principle throughout the world. The descriptions are, of course, a personal view and should only be taken as a guide. The reader is advised to study the original legislation before taking any safety-related decisions.

Most safety legislation has a common theme. Employers and employees are deemed to have a *Duty of Care* to ensure the Health, Safety and Welfare of the employees, visitors and the public. Failure in this duty of care is called *Negligence*. Legislation defines required actions at three levels:

Shall or *Must* are absolute duties which have to be obeyed without regard to cost. If the duty is not feasible the related activity must not take place.

If practicable means the duty must be obeyed if feasible. Cost is not a consideration. If an individual deems the duty not to be feasible, proof of this assertion will be required if an incident occurs.

Reasonably practicable is the trickiest as it requires a balance of risk against cost. In the event of an incident an individual will be required to justify the actions taken.

There is a vast amount of safety legislation with varying degrees of authority. Acts (e.g. the Health and Safety at Work Act (HASWA)), are statutes passed by full parliamentary procedures and are enforced by criminal law. Often acts such as HASWA (called Enabling Acts), are arranged to allow supplementary regulations to be made by the Secretary of State without going through the full parliamentary procedure.

Regulations are introduced under an enabling act. They have the same power and status as acts. Most British safety regulations have been made under the Health and Safety at Work Act 1974.

Approved Codes of Practice (ACOPs) are documents written to define safe working methods and procedures by organizations such as CENELEC and British Standards Institute. They are approved by the Health and Safety Commission. Whilst they are not mandatory (i.e. there can be no prosecution for not following them), failure to follow ACOPs may be viewed as a contributory factor in investigations of an incident.

Codes of Practice are guidance codes provided by trade unions and professional organizations. These do not have the semi-legal status of ACOPs, but contain good advice. Again, though, implementation or otherwise can be given in evidence in court.

In Europe there is a serious attempt to have uniform legislation throughout the EC. At the top level is EC Regulations which override national legislation. Of most relevance are EC Directives which require national laws to implemented.

In Britain the primary legislation is the Health & Safety at Work Act 1974 (HASWA). It is an enabling act, allowing other legislation to be introduced. It is wide ranging and covers everyone involved with work (both employers and employees) or affected by it. In the USA the Occupational Safety and Health Act (OSHA) affords similar protection.

HASWA defines and builds on general duties to avoid all possible hazards, and its main requirement is described in section 2(1) of the act:

> *It shall be the duty of every employer to ensure, so far as is reasonably practicable, the health, safety and welfare at work for his employees*

This duty is extended in later sections to visitors, customers, the general public and (upheld in the courts), even trespassers. The onus of proof of *Reasonably Practicable* lies with the employer in the event of an incident.

Section 2(2) adds more detail by requiring safe plant, safe systems of work, safe use of articles and substances (i.e. handling, storage and transport), safe access and egress routes, safe environment, welfare facilities and adequate information and training.

If an organization has five or more employees it must have a written safety policy defining responsibilities and employees must be aware of its existence and content (section 2(3)) Employers must consult with worker safety representatives

The act is not aimed purely at employers, employees also have duties described in sections 7 and 8 of the act. They are responsible for their own, and other's safety and must co-operate with employers and other people to ensure safety, i.e. they must follow safe working practices. They must not interfere with any safety equipment (e.g. tampering with interlocks on movable guards).

The act defines two authorities and gives them power for the enforcement of the legislation (sections 10–14 and 18–24). The Health and Safety Commission is the more academic of the two, and defines policy, carries out research, develops safety law and disseminates safety information. The Health & Safety Executive (HSE) implements the law by inspection and can enforce the law where failings are found. Breaches of HASWA amount to a indictable offence and the HSE has the power to prosecute the offenders.

The power of HSE inspectors are wide. They can enter premises without invitation and take samples, photographs, documents, etc. Breaches of HASWA amount to a indictable offence and the HSE has the power to prosecute the offenders. People, as well as organisations, may be prosecuted if a safety failing or incident arises because of neglect by a responsible person.

The HSE also has the power to issue notices against an organisation. The first, an Improvement Notice, is given where a fairly minor safety failing is observed. This notice requires the failing to be rectified within a specified period of time. The second, a Prohibition Notice, requires all operations to cease immediately and not restart until the failing is rectified and HSE inspectors withdraw the notice.

It is all but impossible to design a system which is totally and absolutely fail-safe. Modern safety legislation, such as the Six Pack, recognises the need to balance the cost and complexity of the safety system against the likelihood and severity of injury. The procedure, known as *Risk Assessment*, uses common terms with specific definitions:

Hazard	The potential to cause harm.
Risk	A function of the likelihood of the hazard occurring and the severity.
Danger	The risk of injury.

Risk assessment is a legal requirement under most modern legislation, and is covered in detail in, standard prEN1050 '*Principles of Risk Assessment*'.

The first stage is identification of the hazards on the machine or process. This can be done by inspections, audits, study of incidents (near misses) and, for new plant, by investigation at the design stage. Examples of hazards are: impact/crush, snag points leading to entanglement, drawing in, cutting from moving edges, stabbing, shearing (leading to amputation), electrical hazards, temperature hazards (hot and cold), contact with dangerous material and so on. Failure modes should also be considered, using standard methods such as HAZOPS (Hazard & Operability Study, with key words Too much of and Too little of), FMEA (Failure Modes and Effects Analysis) and Fault Tree Analysis.

With the hazards documented the next stage is to assess the risk for each. There is no real definitive method for doing this, as each plant has different levels of operator competence and maintenance standards. A risk assessment, however, needs to be performed and the results and conclusions documented. In the event of an accident, the authorities will ask to see the risk assessment. There are many methods of risk assessment, some quantitative assigning points, and some using broad qualitative judgements.

Whichever method is used there are several factors that need to be considered. The first is the severity of the possible injury. Many sources suggest the following four classifications:

Fatality	One or more deaths.
Major	Non reversible injury, e.g. amputation, loss of sight, disability.
Serious	Reversible but requiring medical attention, e.g. burn, broken joint.
Minor	Small cut, bruise, etc.

The next step is to consider how often people are exposed to the risk. Suggestions here are:

Frequent	Several times per day or shift.
Occasional	Once per day or shift.
Seldom	Less than once per week.

Linked to this is how long the exposure lasts. Is the person exposed to danger for a few seconds per event or (as can occur with major maintenance work), several hours? There may also be a need to consider the number of people who may be at risk; often a factor in petro-chemical plants.

Where the speed of a machine or process is slow, or there is a lengthy and obvious (e.g. noisy) start-up, the exposed person can easily move out of danger in time. There is obviously less risk here than with a silent high speed machine which can operate before the person can move. From studying the machine operation, the probability of injury in the event of failure of the safety system can be assessed as:

Certain, Probable, Possible, Unlikely

From this study, the risk of each activity is classified. This classification will depend on the application. Some sources suggest applying a points scoring scheme to each of the factors above then using the total score to determine *High, Medium* and *Low* risks. Maximum Possible Loss (MPL) for example uses a 50 point scale ranging from 1 for a minor scratch to 50 for a multi-fatality. This is combined with the frequency of the hazardous activity (F) and the probability of injury (again on a 1–50 scale) in the formula:

$$\text{risk rating (RR)} = F \times (\text{MPL} + P)$$

The course of action is then based on the risk rating.

An alternative and simpler (but less detailed approach) uses a table as Figure 8.1 from which the required action can be quickly read.

Likelihood of incident
9. Almost certain
8. Very likely
7. Probable
6. Better than even chance
5. Even chance
4. Less than even chance
3. Improbable
2. Very improbable
1. Almost impossible

Severity of outcome
9. Fatality
8. Permanent total incapacity
7. Permanent severe incapacity
6. Permanent slight incapacity
5. Off work for > 3 weeks but subsequent recovery
4. Off work for 3 days to 3 weeks with full recovery
3. Off work for less than 3 days with full recovery
2. Minor injury, no lost time
1. Trivial injury

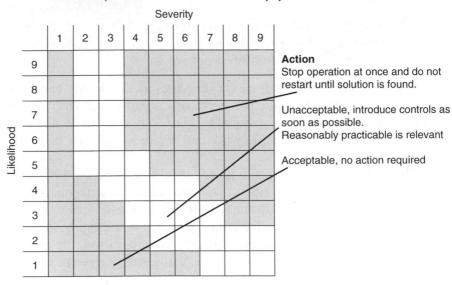

Action
Stop operation at once and do not restart until solution is found.

Unacceptable, introduce controls as soon as possible.
Reasonably practicable is relevant

Acceptable, no action required

Figure 8.1 *A typical risk assessment table. Although this is based on a real application, it should not be applied elsewhere without supporting study and documentation. The main point of a risk assessment is identifying and reducing the risks associated with a specific task*

There is, however, no single definitive method, but the procedure used must suit the application and be documented. The study and reduction of risks is the important aim of the activity.

The final stage is to devise methods of reducing the residual risk to an acceptable level. These methods will include removal of risk by good design (e.g. removal of trap points), reduction of the risk at source (e.g. lowest possible speed and pressures, less hazardous material), containment by guarding, reducing exposure times, provision of personal protective equipment and establishing written safe working procedures which must be followed. The latter implies competent employees and training programs.

There is a vast amount of legislation covering health and safety, and a list is given below of those which are commonly encountered in industry. It is by no means complete, and a fuller description of these, and other, legislation is given in the third edition of the author's *Industrial Control Handbook*. An even more detailed study can be found in *Safety at Work* by John Ridley, both books published by Butterworth-Heinemann.

Commonly Encountered Safety Legislation:
Health & Safety at Work Act 1974 (the prime UK legislation)
Management of Health & Safety at Work Regulations 1992
Provision & Use of Work Equipment Regulations 1992 (PUWER)
Manual Handling Regulations 1992
Workplace Health, Safety & Welfare Regulations 1992
Personal Protective Equipment Regulations 1992
Display Screen Equipment Regulations 1992
   (*the previous six regulations are based on EC directives and are
   known collectively as 'the six pack'*)
Reporting of Injuries, Diseases & Dangerous Occurrences
   Regulations (RIDDOR) 1995
Construction (Design & Management) Regulations (CDM) 1994
   Electricity at Work Regulations (1990)
Control of Substances Hazardous to Health (COSHH) 1989
Noise at Work Regulations 1989
Ionising Radiation Regulations 1985
Safety Signs & Signals Regulations 1996
Highly Flammable Liquids & Liquefied Petroleum Gas
   Regulations 1972
Fire Precautions Act 1971
Safety Representative & Safety Committee Regulations 1977
Health & Safety Consultation with Employees Regulations 1996
Health & Safety (First Aid) Regulations 1981
Pressure Systems & Transportable Gas Containers Regulations
   1989

As hydraulic systems are nowadays invariably linked to Programmable Controllers (PLCs), the reader should also consult the occasional paper OP2 'Microprocessors in Industry' published by the HSE in 1981 and the two later booklets 'Programmable

Electronics Systems in Safety Related Applications', Book 1, an Introductory Guide and Book 2, General Technical Guidelines both published in 1987.

Electrical systems are generally recognised as being potentially lethal, and all organisations must, by law, have procedures for isolation of equipment, permits to work, safety notices and defined safe-working practices. Hydraulic and pneumatic systems are no less dangerous; but tend to be approached in a far more carefree manner. High pressure air or oil released suddenly can reach an explosive velocity and can easily maim, blind or kill. Unexpected movement of components such as cylinders can trap and crush limbs. Spilt hydraulic oil is very slippery, possibly leading to falls and injury. It follows that hydraulic and pneumatic systems should be treated with respect and maintained or repaired under well defined procedures and safe-working practices as rigorous as those applied to electrical equipment.

Some particular points of note are:

- before doing *anything,* think of the implications of what you are about to do, and make sure anyone who could be affected knows of your intentions. Do not rush in, instead, *think*;

- anything that can move with changes in pressure as a result of your actions should be mechanically secured or guarded. Particular care should be taken with suspended loads. Remember that fail open valves will turn *on* when the system is de-pressurised;

- never disconnect pressurised lines or components. Isolate and lock-off relevant legs or de-pressurise the whole system (depending on the application). Apply safety notices to inhibit operation by other people. Ideally the pump or compressor should be isolated and locked off at its MCC. Ensure accumulators in a hydraulic system are fully blown down. Even then, make the first disconnection circumspectly;

- in hydraulic systems, make prior arrangements to catch oil spillage (from a pipe-replacement, say). Have containers, rags and so on, ready and, as far as is possible, keep spillage off the floor. Clean up any spilt oil before leaving;

- where there is any electrical interface to a pneumatic or hydraulic system (eg, solenoids, pressure switches, limit switches) the control circuits should be isolated, not only to remove the risk of electric shock, but also to reduce the possibility of fire or accidental initiation of some electrical control sequence. Again, *think* how things interact;

- after the work is completed, leave the area tidy and clean. Ensure people know that things are about to move again. Check there is no one in dangerous areas and sign-off all applied electrical, pneumatic or hydraulic isolation permits to work. Check for leaks and correct operation;

- many components contain springs under pressure. If released in an uncontrolled manner these can fly out at high speed, causing severe injury. Springs should be released with care. In many cases manufacturers supply special tools to contain the spring and allow gradual and safe decompression.

Cleanliness

Most hydraulic or pneumatic faults are caused by dirt. Very small particles nick seals, abrade surfaces, block orifices and cause valve spools to jam. In hydraulic and pneumatic systems cleanliness is next to Godliness. Dismantling a valve in an area covered in swarf or wiping the spool on an old rag kept in an overall pocket does more harm than good.

Ideally components should not be dismantled in the usual dirty conditions found on site, but returned to a clean workshop equipped with metal-topped benches. Too often one bench is used also for general mechanical work: it needs little imagination to envisage the harm metal filings can do inside a pneumatic or hydraulic system.

Components and hoses come from manufacturers with all orifices sealed with plastic plugs to prevent dirt ingress during transit. These should be left in during storage and only removed at the last possible moment.

Filters exist to remove dirt particles, but only work until they are clogged. A dirty filter bypasses air or fluid, and can even make matters worse by holding dirt particles then releasing them as one large collection. Filters should be regularly checked and cleaned or changed (depending on the design) when required.

Oil condition in a hydraulic system is also crucial in maintaining reliability. Oil which is dirty, oxidised or contaminated with water forms a sticky gummy sludge, which blocks small orifices and causes pilot spools to jam. Oil condition should be regularly checked and suspect oil changed before problems develop.

Fault-finding instruments

Electrical fault-finding is generally based on measurements of voltage, current or (less often) resistance at critical points in the circuit. Of these, voltage is easier to measure than current unless ammeters or shunts have been built into the circuit, and resistance measurement usually requires the circuit to be powered-down and the device under test disconnected to avoid sneak paths. An electronic circuit is given in Figure 8.2. This converts a voltage input V_i to a current signal I, where $I = V_i/R$. Such a circuit is commonly used to transmit an instrumentation signal through a noisy environment. A typical checking procedure could be:

Voltage checks;	A	(input signal)
	B and C	(amplifier ± 15 V supply)
	D	(return voltage should equal A)
	E	(across load, 15 V indicates open circuit load, 0 V indicates short-circuit load).

Followed by:

Current checks;	X, Y	(X should equal Y and both equal A/R)
Resistance checks;	F, G	(for open- or short-circuit load or resistance).

In pneumatic or hydraulic systems, pressure measurement is equivalent to electrical voltage measurement, while flow measurement is equivalent to current measurement. There is no direct simple measurement equivalent to electrical resistance. Pressure tests and (to a lesser extent) flow tests thus form the bases of fault-finding in pneumatic or hydraulic systems.

There is however, a major difference in the ease of access. Electrical systems abound with potential test points; a voltage probe can be placed on practically any terminal or any component, and (with a little more trouble) a circuit can be broken to allow current measurements to be made.

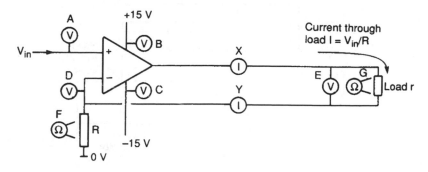

Figure 8.2   *Test measurement points on an electronic circuit*

In fluid systems, oil or gas is contained in pipes or hoses, and measurements can only be made at test points which have been built-in as part of the original design. Test points can be plumbed in on an ad hoc basis but this carries the dangers of introducing dirt from cutting or welding, and in hydraulic systems any air introduced will need to be bled out. The designer should, therefore, carefully consider how faults in the system can be located, and provide the necessary test points as part of the initial design.

By far the most common technique is a built-in rotary pressure select switch, as shown in Figure 8.3, which allows pressure from various strategic locations to be read centrally. An alternative technique uses quick-release connections, allowing a portable pressure meter to be carried around the system and plugged in where required.

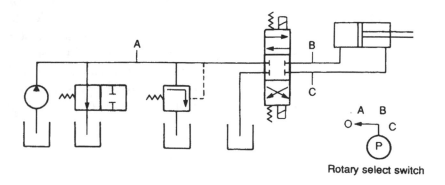

Figure 8.3   *The commonest hydraulic and pneumatic test system, a rotary select switch*

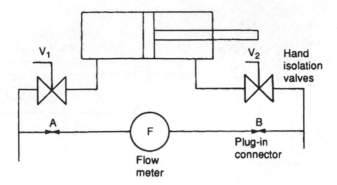

Figure 8.4 *Checking flow available at an actuator*

Flow measurement is more difficult, as the basic flow transducer needs to be built-in. Portable flow meters can be used, as shown in Figure 8.4, where the flow available for a cylinder is checked by closing hand valves V_1 and V_2 while connecting a flow meter between quick-release connections A and B.

The UCC System 22 is an invaluable three in one 'plumbed-in' test instrument which provides measurement of pressure, flow and temperature at the installed point with a plug in test meter. The inclusion of such a device should be provided immediately after every pump (but before the first relief valve) to allow pump delivery to be checked and at crucial points such as the pressure lines to a critical cylinder or motor. Remember, with hydraulics and pneumatics the test points have to be designed in.

An indicator in the plug of a solenoid valve will show voltage is arriving at the solenoid (see Figure 8.9) but this is not a fool proof indication that the solenoid itself is operating. The coil may, for example, be open circuit or there is a loose connection inside the plug. RS components sell a very cheap and useful solenoid tester (part number 214–338), which illuminates when held in a strong magnetic field. About the size of a fountain pen it can be touched onto the body of a solenoid to see if the solenoid really is being energized.

Fault-finding

Fault-finding is often performed in a random and haphazard manner, leading to items being changed for no systematic reason beyond 'Fred got it working this way last time'. Such an approach may work eventually (when every component has been changed!) but it is hardly the quickest, or cheapest, way of getting a faulty system back into production. In many cases more harm than good results, both with introduction of dirt into the system, and from ill advised 'here's a control adjustment; let's twiddle it and see if that makes any difference' approach. There must be a better way.

There are three maintenance levels. First line maintenance is concerned with getting faulty plant running again. When the cause of a fault is found, first line staff have the choice of effecting a first line, on site, repair (by replacing a failed seal, say) or changing the complete faulty unit for a spare. This decision is based on cost, time, availability of spares, technical ability of staff, the environment on site and company policy.

Second line maintenance is concerned with repair to complete units changed by first line maintenance staff. It should be performed in clean and well-equipped workshops. Work is usually well-defined and is often a case of following manufacturers' manuals.

The final level is simply the return of equipment for repair by manufacturer. The level at which this is needed is determined by the complexity of equipment, ability of one's staff, cost and the turn-round time offered by the manufacturer.

Of these three levels; first line maintenance is hardest as work is ill-defined, pressures from production staff are great and the responsibility high. Unfortunately, it is too often seen as a necessary evil.

Fault-finding is, somewhat simplistically, represented by Figure 8.5. All the evidence on the fault gathered so far is evaluated, and possible causes considered. The simplest test to reduce the number of possibilities is then performed and the cycle repeated until the fault is found.

The final steps in Figure 8.5 are concerned with fault recording and fault analysis. Any shift crew (which performs almost all the first line repairs) only sees one quarter of all faults. The fault recording and analysis process shows if there is any recurring pattern in faults, indicating a design or application problem. Used diplomatically, the records may also indicate shortcomings in crews' knowledge and a need for training.

Modern plants tend to be both complex and reliable. This means that a maintenance crew often sees a plant in detail for the first time when the first fault occurs. (Ideally, of course, crews should be involved at installation and commissioning stages – but that is another story!). It is impossible to retain the layout of all bar the

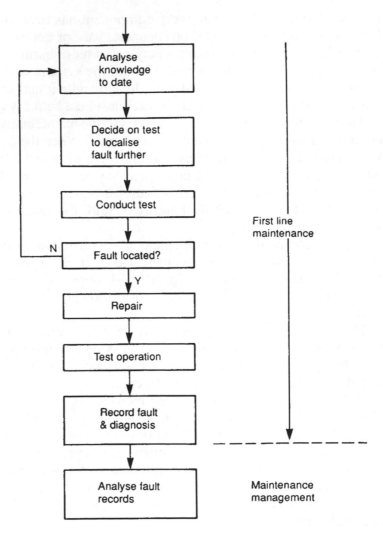

Figure 8.5 *Fault-finding process*

simplest of systems in the mind, so it is essential to have schematic diagrams readily available.

Equally important, readings at each test point should be documented when the system is working correctly. It is not much use to know pressure at TP_3 is 15 bar, the motor draws 75 A or flow to rotary actuator C is 1500 l min^{-1} under fault conditions, without knowing what the normal readings are.

It can often be difficult to decide what a fault really is; usually the only information is simply 'the Firkling Machine is not working'. The first diagnostic step is, therefore, to establish what is *really* wrong – whether there is one fault or several from a common cause. A quick visual and manual check should be made for any obvious aberrations; noise, vibration, heat, leaks, unusual motor current.

From maintenance records it should be possible to see if any recent work has been done or if this is a recurring fault. Recent work is always suspect – particularly if the unit has not been used since the work was done. Some points to check if work has recently been performed are:

• whether the correct units were fitted. Stores departments are not infallible and lookalike units may have been fitted in error. Non-return valves can sometimes be fitted the wrong way round;

• whether all handvalves are correctly open or shut. Many systems are built with standby pumps or compressors with manual change over. These are (in the author's experience) a constant source of trouble after a changeover (one invariable characteristic seems to be one less valve handle than there are valves!). Valves can also creep open. Figure 8.6 shows a common fault situation with two hydraulic units, one in use and one standby. If any of the hand isolation valves V_1 to V_4 are set incorrectly open on the main or standby units, flow from the duty unit returns direct to tank via the centre position of the standby directional valve and the actuator will not move;

• after electrical work, check the direction of rotation of the pump or compressor. Most only operate in one direction, usually defined with an arrow on the casing, and may even be damaged by prolonged reverse running;

• have any adjustments been 'twiddled' or not set correctly after an item has been changed? On many directional valves, for example, the speeds of operation from pilot to main spool can be set by Allen key adjustments. If these are maladjusted, the main spool may not move at all.

If no recent work has been done, and these quick checks do not locate a fault, it is time to start the fault-finding routine of Figure 8.5. One advantage of pneumatic systems is their natural break into distinct portions: (1) a supply portion up to and including the receiver and (2) one or more application portions after the receiver.

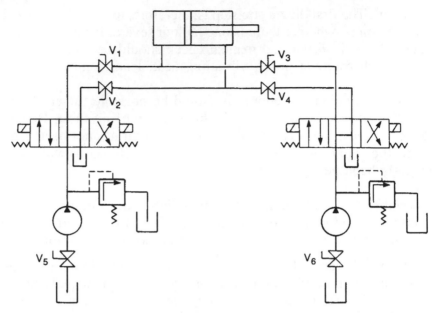

Figure 8.6 *A common source of trouble; a main/standby system with hand isolation valves. Wrong setting of valves leads to many obscure faults*

The pressure gauge on the receiver allows a natural fault-finding split.

Problems generally fall into three types; a lack of force, low speed (or no speed), or erratic operation. Lack of force or no movement is generally a pressure-related fault. Low speed arises from a flow fault. Erratic operation can arise from sticking valves or from air in a hydraulic system.

Usually pressure monitoring is much easier than flow monitoring but is often misunderstood. A typical example of fault-finding using pressure test points is given in Figure 8.7. Up to time A the system unloads via the solenoid-operated unloading valve V_1. When valve V_1 energises, pressure rises to the setting of the relief valve V_2. At time C, directional valve V_3 calls for the cylinder to extend. Pressure *falls* as the cylinder accelerates, until the cylinder is moving at constant speed when P=F/A. At time D, the cylinder reaches the end of travel, and the pressure rises back to the setting of the relief valve. Directional valve V_3 de-energises at E. Note the low pressures in return line test points.

A similar retract stroke takes place during time F to I. The pressure between G and H is lower than between C and D, because fric-

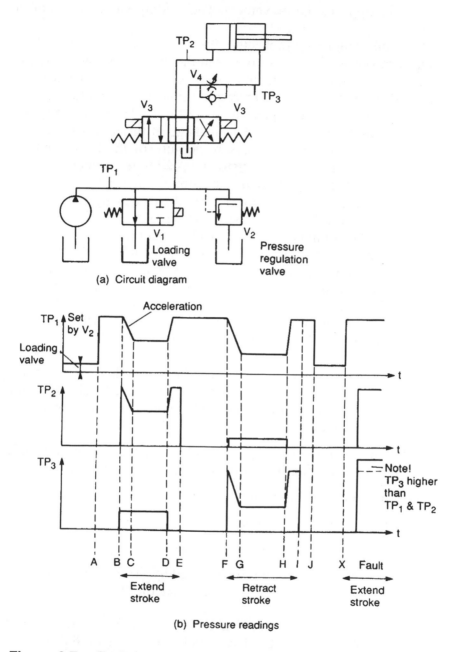

(a) Circuit diagram

(b) Pressure readings

Figure 8.7 *Fault-finding with pressure test points*

tion alone opposes the movement. The loading valve comes off at time J.

It is important to monitor return line pressure. A fault exists from time X onwards; the return line from the cylinder is blocked, possibly because the spool in the meter-out flow control valve has jammed, allowing no fluid to return. At time Y the directional valve operates causing a rise in pressure on test point TP_2 to the setting of the relief valve. Because of the blockage in the return line, point TP_3 also rises to a *higher* pressure because of the lower annulus area on the return side of the piston ($P_1A=P_2a$, remember!).

Pressure is therefore a good indication of what is going on in a system, the pressure being the *lowest* demanded by the loading/unloading valves, the relief valve(s) or the load itself.

A hydraulic pump is a positive displacement device (see Chapter 2). This has useful implications when fault-finding. If a pump is working its flow *must* be getting back to tank via some route. If it does not the pressure will rise and the oil will eventually go everywhere! Tracking the oil flow route by as simple a method as following warm pipes by hand can sometimes indicate what is wrong.

Remember these basic facts for fault finding:

- Knowing if the pump is delivering fluid is vital. If there is not a UCC system 22 or similar flow sensor immediately after the pump but before the first relief valve consider installing one as soon as possible

- Hydraulic pumps are invariably positive displacement pumps. If a pump is delivering fluid it must be going somewhere.

- Acceleration is determined by pressure

- Force is determined by pressure

- Velocity (speed) is determined by flow

- The pressure at any point is determined by the *lowest* pressure the system can provide under the current conditions.

The interface with the electrical control can cause confusion. The control sequence should be clearly understood. Figure 8.8 shows a typical electrical/hydraulic scheme used to build a tight pack of objects. An object is placed onto the skid, and its presence noted by a proximity detector connected as an input to a programmable controller (PLC).

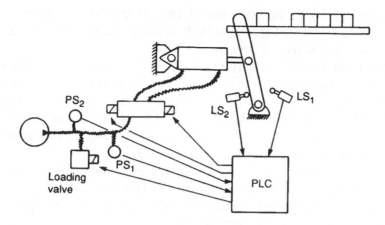

Figure 8.8 *A typical sequencing application*

When the PLC sees an object, it energises the loading valve, and causes the cylinder to extend. The cylinder extends until the front limit switch LS_1 makes (for the first few objects) *or* the pressure switch PS_2 makes (indicating a full stalled pack) *or* timeout (indicating some form of fault). The cylinder then returns to the back limit LS_2 *or* a timeout (again indicating a fault) when the loading valve is de-energised. The PLC also monitors pump action via pressure switch PS_1, which is made whenever the loading valve is energised.

A knowledge of the complete system, both electrical and hydraulic, is required to fault-find on this application. Fault-finding involves checking the sequence by monitoring the state of electrical outputs to solenoids and inputs from limit switches.

All solenoid valves should have an indicator in the plug tops to allow electrical signals to be observed local to the valves. Indicator blocks which fit between plug and valve are available for retro fitting onto systems without this useful feature. It should be remembered, though, that indications purely show electrical voltage is present – it does not, for example, identify an open circuit solenoid coil.

Solenoids can operate on AC (usually 110 V AC) or DC (usually 24 V DC). DC solenoids have totally different operating characteristics. An AC solenoid has a very high inrush current producing a high initial force on the pilot spool. As the spool moves in, the inductance of the coil rises and current falls to a low holding current (and a low force on the pilot spool). If the pilot spool jams the current remains high, causing the protection fuse or breaker to open or the solenoid coil to burn out if the protection is inadequate. Operation of a 110 volt solenoid system with cold oil is best undertaken with a pocketful of fuses.

The current in a DC solenoid is determined by the coil resistance and does not change with pilot spool position. The solenoid does not, therefore, give the same 'punch' to a stiff spool but will not burn out if the spool jams. Current in a DC solenoid also tends to be higher requiring larger size cables, particularly if a common return line is used from a block of solenoids.

A useful monitoring device is the through-connector with an integral current indicator shown in Figure 8.9. This *does* give indication of an open circuit coil and combined with the indicator on the plug top can help find most electrical faults.

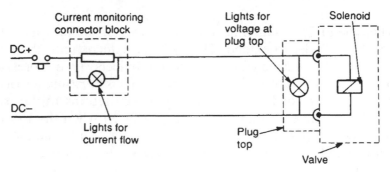

Figure 8.9   *Monitoring a DC solenoid*

On most electrically-operated valves the pilot spool can be operated manually by pushing the spool directly with a rod (welding rod is ideal!). Electrical signals should, however, be disabled when operating valves manually, as pushing in the opposite direction to the solenoid can cause the coil to burn out.

Designers of a system can simplify maintenance by building in a fault-finding methodology from the start. This often takes the form of a flowchart. Figure 8.10 shows a typical system, which can be diagnosed by following the flowchart of Figure 8.11. Such charts cannot solve every problem, but can assist with the majority of common faults. If transducers can be fitted to allow the system to be monitored by a computer or programmable controller, Figure 8.11 could form the basis of a computer-based expert system.

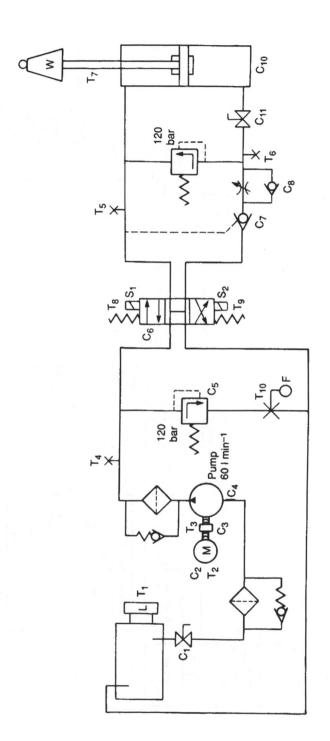

Figure 8.10 *Hydraulic circuit for diagnostic chart*

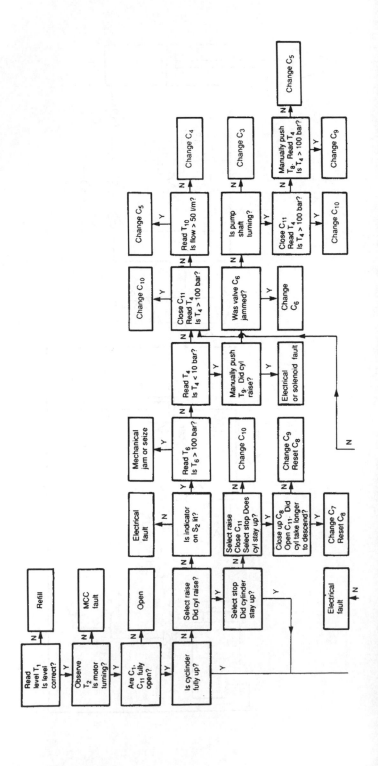

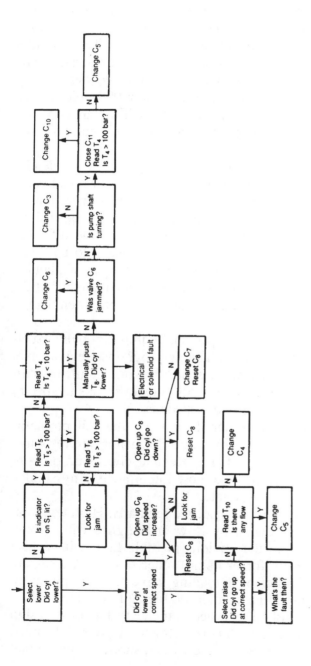

Figure 8.11 *Fault-finding flowchart for circuit of Figure 8.10*

Preventive maintenance

Many production people think a maintenance department exists purely to repair faults as they occur (the common image being a team sitting in the workshop waiting for the 'phone to ring). The most important part of a maintenance department's responsibility, however, is performing routine planned maintenance. This provides regular servicing of equipment, checks for correct operation and identifies potential faults – which can be corrected before they interrupt production. A personal analogy is the 6,000 mile service for motor cars. As an often overlooked side benefit. planned maintenance trains the maintenance craftsmen in the operation and layout of the plant for which they are responsible.

A planned maintenance schedule can be based on a calendar basis (work done daily, weekly, monthly and so on) or on an operation based schedule (work done after so many hours operation, or so many cycles) with time run or number of cycles recorded by control equipment. Different parts of the system may have differing maintenance schedules. Identifying what work needs to be done, and the basis of the schedule for each item is the art of planned maintenance. It depends heavily on the nature of the plant; air filters in a dust filled steel works say, require checking more often than in a clean food factory.

With the advent of the desk-top personal computer many excellent computer-based maintenance planning programs are available. These produce fully detailed work schedules on a shift-by-shift basis, and flag urgent work. The user still, however, has to specify the work to be done and the basis of schedules.

In hydraulic systems it is generally thought that oil problems (level in the tank, contamination by dirt, air or water) are responsible for around three-quarters of faults. Regular checks on oil condition and level are therefore of utmost importance. Any sudden change in level should be investigated.

Oil temperature should also be checked regularly. High temperatures arise from heat produced by flow discharging with a high pressure drop. Apart from the obvious possible fault with a heat exchanger (no water flow for example) other possible causes are incorrect operation of relief or unloading valves (ie, the pump on load continuously) internal leakage or too high a fluid viscosity.

System pressure should be recorded and checked against design values. Deviations can indicate maladjustment or potential faults. Too high a pressure setting wastes energy and shortens operational life. Too low a pressure setting may cause relief valves to operate at pressures below that needed by actuators, leading to no movement. Pressure deviation can also indicate developing faults outside the system. The fouling of a component moved by an actuator, for example, may cause a rise of pressure which can be observed before a failure occurs.

Motor currents drawn by pumps and compressors should also be checked both in working and unloading states (ideally, indication of motor currents should be available on a panel local to the motor). Changes in current can indicate a motor is working harder (or less) than normal.

Filters are of prime importance in both hydraulic and pneumatic systems. The state of most hydraulic filters is shown by a differential pressure indicator connected across the filter element. Obviously filters should be changed before they become blocked. Inlet air filters on pneumatic systems also need regular cleaning (but *not* with flammable fluids such as petrol or paraffin). A record should be kept of filter changes.

Many checks are simple and require no special tools or instruments. Visual checks should be made for leaks in hydraulic systems (air leaks in pneumatic systems generally can be detected from the noise they make!). Pipe runs and hosing should be visually checked for impact damage and to ensure all supports are intact and secure. Connections subject to vibration should be examined for tightness and strain. It is not unknown for devices such as pumps and compressors to 'walk' across the floor dragging their piping with them.

Where the device examined follows a sequence, the operation should be checked to ensure all ancillary devices, such as limit switches, are operating. The time to perform sequences may be worth recording as a lengthening of sequence times may indicate a possible developing fault due to, say leakage in a cylinder.

Actuators have their own maintenance requirements given in manufacturers' manuals. Seals and bushing in cylinders, for example, require regular checking and replacement if damaged. Cylinder rods should be examined for score marks which can indicate dust ingress. Actuators which move infrequently under normal duty can be operated to check they still work (and also to help lubricate the seals).

Treat leaks from around the rods of cylinders with urgency. If oil is leaking out round the neck seal on the extend stroke, dirt is being drawn into the system on the return stroke and a minor leak can soon turn into a major system failure.

Pneumatic preventive maintenance is very similar to hydraulic maintenance (although obviously there is no hydraulic oil to check). Other points such as piping, filters, fittings, sequences and so on need checking in the same way.

Compressors have their own maintenance requirements Many are belt driven, and require belt condition and tension to be checked at regular intervals. Crankcase oil level and the air breather should also be checked.

The compressor is normally sized for the original capacity plus some reserve for future additions. A compressor will thus start life on a low duty cycle, which increases as further loads are added. When compressor capacity is reached, the compressor will be on 100% duty cycle. Any additional load results in a fall of system pressure in the receiver. Leaks also cause a rise in compressor duty cycle, as will any loss of compressor efficiency. Duty cycle of the compressor thus gives a good indication of the health and reserve capabilities of the systems.

Compressor efficiency is determined largely by the condition of valves, piston rings and similar components subject to friction wear. These should be examined at intervals given in manufacturers' instruction manuals.

Other common pneumatic maintenance checks are validation of safety valve operation on the receiver, replenishment of oil in the air lubrication and drainage of water from air dryers.

Index